DISCOVERING
ASTRONOMY

Cover picture: A close-up view of Saturn taken by the camera on board Voyager 2.

Opposite: In this spectacular photograph taken by Voyager 1, a volcano erupts on the surface of Io, one of the moons of Jupiter. The colors are due to sulfur on the surface of Io.

DISCOVERING
ASTRONOMY
by Jacqueline & Simon Mitton

Published by

STONEHENGE

in association with

The American Museum of Natural History

The authors

Simon Mitton studied Physics at Oxford and is an MA of the universities of Oxford and Cambridge. He studied for his PhD in radio astronomy under Sir Martin Ryle, the Nobel Laureate and Astronomer Royal. Dr Mitton is the Editor-in-Chief of *The Cambridge Encyclopedia of Astronomy* and author of *Exploring the Galaxies*. He has also published a number of scientific papers, and written many popular articles on current trends in astronomy.

Jacqueline Mitton studied Physics at Oxford and studied for her PhD at the Cambridge Observatories. For several years she has been doing research work at the Cambridge Institute of Astronomy. Her special interest is spectroscopy, in particular the light analysis of the brightest stars.

The consultants

Dr Malcolm Longair is Regius Professor of Astronomy at the University of Edinburgh, Astronomer Royal for Scotland and Director of the Royal Observatory, Edinburgh. He took his PhD at the University of Cambridge and has been a University Lecturer in the Department of Physics at the University of Cambridge, visiting Assistant Professor of Radioastronomy at the California Institute of Technology, and visiting Professor of Astronomy at the Institute of Advanced Study, Princeton.

Dr. Thomas D. Nicholson is Director at The American Museum of Natural History at New York. He was formerly astronomer and chairman at the Museum's Department of Astronomy, the Hayden Planetarium, and he has written, taught and lectured in astronomy extensively. A master mariner, he is also a recognized authority on celestial navigation.

The American Museum of Natural History

Stonehenge Press wishes to extend particular thanks to Dr. Thomas D. Nicholson, Director of the Museum, and Mr. David D. Ryus, Vice President, for their counsel and assistance in creating this volume.

Stonehenge Press Inc.:
Publisher: John Canova
Editor: Ezra Bowen
Deputy Editor: Carolyn Tasker

Trewin Copplestone Books Ltd:
Editorial Consultant: James Clark
Managing Editor: Barbara Horn

Created, designed and produced by Trewin Copplestone Books Ltd, London.

© Trewin Copplestone Books Ltd, 1979, 1982
All rights reserved. No part of this book may be reproduced or utilized in any form or by any means, electronic or mechanical, including photocopying, recording or by any information storage or retrieval system without permission in writing from Trewin Copplestone Books Ltd, London.

Library of Congress Card Number: 82-50149
Printed in U.S.A. by Rand McNally & Co.
First printing
ISBN 0-86706-016-6
ISBN 0-86706-063-8 (lib. bdg.)
ISBN 0-86706-032-8 (retail ed.)
Set in Monophoto Rockwell Light by
SX Composing Ltd, Rayleigh, Essex, England
Separation by Gilchrist Bros. Ltd, Leeds, England
Printed in U.S.A. by Rand NcNally & Co.

Contents

The World of Astronomy

The science of astronomy includes the study of the planets, stars and galaxies. The Earth is a planet that orbits, or traces a path, around the Sun. A planet, such as Earth, shines only by the light that it reflects, whereas a star, such as the Sun, generates its own light. The family of nine planets associated with the Sun is known as the solar system. Many of these planets have smaller objects, known as satellites, orbiting them. The Moon, for example, is the Earth's natural satellite, and also shines by the light it reflects from the Sun. The planets of the inner solar system are Mercury, Venus, Earth and Mars, sometimes called the terrestrial planets because they are small and made of rock. In the outer solar system are four planets that are much larger than Earth – Jupiter, Saturn, Neptune, Uranus, and a further small planet, Pluto.

The Sun is an ordinary star, similar to billions of other stars in the Universe. It shines by means of the energy released in nuclear reactions that take place in its innermost regions, and that are similar to nuclear processes in hydrogen bombs. The Sun and its neighboring stars are just a tiny part of a gigantic family of stars known as the Galaxy, or Milky Way. This family has about 100,000,000,000 members. Its size is so great that a radio signal traveling at the speed of light, 186,000 miles per second, would take 100,000 years or so to cross from one side to the other. If this imaginary radio signal traveled out beyond the Galaxy, it would require a further two million years to reach the nearest large galaxy, and around ten billion years to encounter the farthest galaxies that have been glimpsed through astronomers' telescopes.

Beyond the limits of the Milky Way, every step must be measured in millions of light years. A light year is the distance light travels in one year, almost six trillion miles. Two million light years past the edge of the Milky Way is another huge galaxy, the Andromeda; ten million light years farther out two dozen large galaxies come into view. At a distance of fifty million light years are a whole cluster of galaxies milling around as one family. Beyond, countless billions of other galaxies stretch out of sight to the far bounds of the Universe.

From Earth, a tiny planet orbiting an ordinary star, only parts of this unimaginably vast Universe of stars and galaxies are visible. The astronomer's task is to understand the nature of the Universe and the things in it. This book is a basic guide to astronomy, intended for everyone who wants to know what astronomers do and what they have already discovered about the planets, stars and galaxies.

The Nightly Parade of the Stars

After the Sun sets on a clear day, pinpoints of starlight gradually become visible in the darkening sky. The stars are there in the day, too, but it is impossible to see them because of the brilliance of the daytime sky. As the Earth's rotation carries the stargazer into darkness, the nightly parade of stars begins. In the same way that the Sun rises in the east and sets in the west, stars seem to rise and set as the Earth spins around. While some stars sink out of view below the western horizon, new ones are rising in the east. By dawn, the starry sky has changed considerably from its appearance at dusk.

For people living north of the Equator, one star hardly moves in the sky: the North Star, or Polaris. This star has, by chance, a rather special position in the sky, for it is almost directly over the Earth's North Pole. The Earth's rotation axis points out into space toward a reference point designated by astronomers as the north pole of the sky. The North Star is very close to the north pole of the sky. There is also a south pole of the sky in the exact opposite direction in space. No bright star lies close to the south pole of the sky.

Stars close to the Pole never rise and set but are always above the horizon. These are called circumpolar stars. As the Earth turns on its axis, the stars trace circles through the sky, around the Poles. An interesting way to show the paths the stars follow during the night is to take a time exposure photograph. A camera is set pointing at the sky with the shutter open. The photograph shows the circular paths of the stars.

The number of circumpolar stars visible from any particular place depends on its latitude on the Earth. Latitude is measured in angular degrees north and south of the Equator. Since the North Star is almost over the Earth's North Pole, an observer there sees it right overhead. All the other stars seem to move in circles around the North Star, none of them ever rising or setting from that point of view. Conversely, to an ob-

The height of the North Star in the sky depends on the latitude of the observer. Within the Arctic Circle (top) the North Star is nearly overhead. In the middle latitudes it is lower, and near the Equator (bottom) the star is close to the horizon.

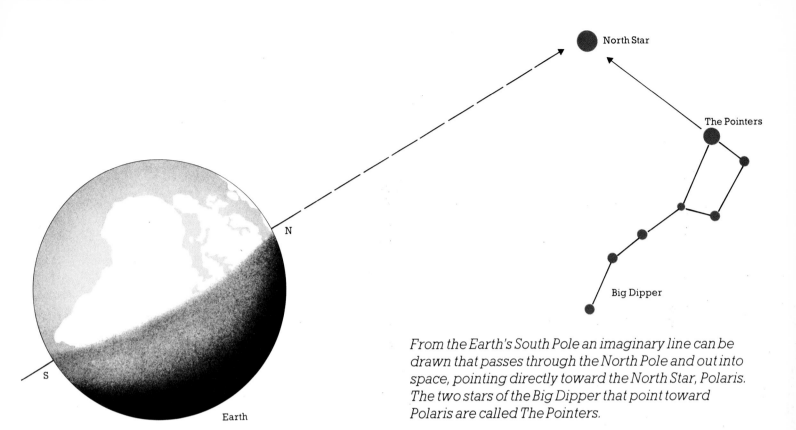

From the Earth's South Pole an imaginary line can be drawn that passes through the North Pole and out into space, pointing directly toward the North Star, Polaris. The two stars of the Big Dipper that point toward Polaris are called The Pointers.

server on the Equator, the North Star is on the northern horizon. No stars are circumpolar; all stars rise and set. If the observer travels north from the Equator, the North Star will look higher and higher in the sky.

Just as the Sun's position can be used to tell the time by day, the stars make a night time clock. The famous group of stars called the Big Dipper is circumpolar for people in North America and Europe. Two of its stars form a line that points to the North Star. This line of stars is like the hour hand of a gigantic 24-hour clock in the sky. During the course of a day, the Big Dipper makes a whole circle around the North Star.

The total number of stars that can be seen by the unaided eye from both the northern and southern hemispheres is around nine thousand.

The time-exposure photograph, at right, shows the circular paths that are traced by stars on the sky as the Earth spins. This is a picture of the southern sky as seen from Australia, above the dome of the 150-inch Anglo-Australian Telescope.

Day and Night

Every day the Sun rises over the eastern horizon. It climbs higher in the sky, following a curved path. In the evening it sets in the west and nightfall soon follows. All animals and plants are influenced by the daily cycle. Some animals rest by day and seek their food at night. Others, including humans, do the opposite. The flowers of many plants open in daylight and close at night. But what is the explanation of "day" and "night"?

In ancient times, people thought that the Sun itself actually traveled across the heavens. The Egyptians, for example, had their sun-god, Re, who daily rode a fiery chariot across the skies. Today people know that it is not the Sun that moves, but the Earth. Earth is like a spinning ball, turning from west to east. One day's rotation makes the Sun appear to move across the sky. The side of the Earth facing the Sun has day. At the same time, on the half of the Earth facing away from the Sun, it is night.

Marking the Sun's progress across the heavens by day was one of the earliest methods used for recording time. A simple way of reading time from the Sun is to use a sundial. A shadow stick, called a gnomon, marks out the hours by the position where its shadow lies. The shadow slowly moves all the time as the Sun follows its course through the sky. Before clocks were invented, most people had to rely on the Sun for keeping time.

It is much easier, of course, to use a watch or clock. Accurate clocks in special laboratories give an exact reading of the time, precise to millionths of a second. But who keeps a check on the master clocks? Astronomers do this by observing the positions of the Sun and stars. If all the world's clocks stopped at once, astronomers could reset them to the correct time.

However, if everyone took the time directly from the Sun, a serious problem would arise. Sun time readings differ for places with different longitudes. Longitude is measured in angular degrees east and west of Greenwich, England. For each 15° of longitude, the Sun time changes by a whole hour. All the people living in one country or area need to agree on just one time to follow if there is not to be a great

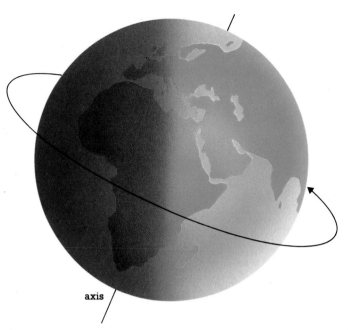

axis

The Sun illuminates only half of the Earth at a time. As the Earth rotates on its axis, from west to east, regions in darkness move into the sunlight. Above, the Sun is rising over eastern Africa and western Europe. Later on, as shown below, dawn breaks over the coast of Labrador on the North American continent.

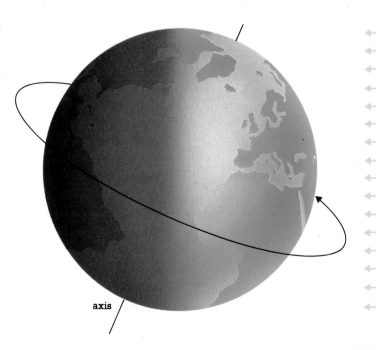

axis

deal of confusion! The time taken directly from the Sun is called local time, whereas the time agreed for a whole country or area is termed standard, or civil time. Standard time is usually chosen to be close to the local time for most of the country.

The entire surface of the Earth is divided into time zones. Each zone is roughly 15° of longitude, but the boundaries are not straight. Instead, they follow the borders between countries or states. Crossing a country or state border, travelers may have to reset the time of their watches. Large countries, such as the United States, need several time zones. Supersonic airliners can travel faster than the speed of the Earth's rotation. This gives travelers going west the strange sensation of arriving at their destination "before" starting on their journey!

Sometimes the governments of countries decide to change standard time by one or even two hours. When this happens, standard time is no longer close to local time. The time change is done to make better use of daylight so that artificial lights are needed less.

It is the night that is of supreme importance to astronomers. No stars can be seen in the daytime, although the planet Venus may sometimes be spotted by a keen-sighted observer. Except for studying the Sun itself, serious astronomy can only begin after sundown when the stars and planets become clearly visible.

An early instrument for recording time is the sundial. The shadow stick, or gnomon, shows the hours according to the position of the Sun at any given time of day. A simple sundial will indicate the time correct to about 15 minutes. This fine dial is in Kew Gardens, London.

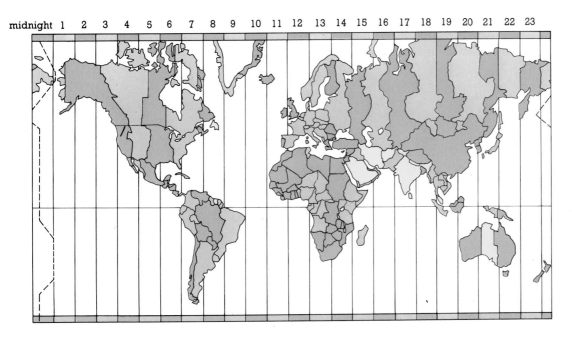

midnight 1 2 3 4 5 6 7 8 9 10 11 12 13 14 15 16 17 18 19 20 21 22 23

The map shows the time zones throughout the world. Areas where the time differs from that in an adjacent zones by less than one hour are indicated in yellow.

The Earth's Yearly Journey

The planet Earth is a member of the solar system. It is one in a family of nine major planets circling the Sun. Each year the Earth travels one complete circuit of its path, or orbit, around the Sun. On the round trip, Earth and everything on it is speeding along at 70,000 miles per hour. People do not feel any sensation of the high-speed motion because they keep moving smoothly at the same rate.

The Earth takes a year to complete its circuit of the Sun. At different seasons the night time side of the Earth looks out at the stars in different directions. Visible in summer and fall (left to right, below) are the constellations Aries, Pisces, Aquarius, Capricornus, Sagittarius, Scorpius and Libra.

What keeps the Earth on its path round the Sun? Although the Sun is an average of 93,000,000 miles away from the Earth, there is a strong gravitational pull between the two bodies. This gravitational force attracts the Earth toward the Sun, but the Earth's speed in its orbit prevents it from falling in toward the Sun. Instead it orbits the Sun. The Earth's orbit is very nearly a circle. It is slightly elongated – a shape called an ellipse – and the Sun is nearer to one side of the Earth's orbit. The Earth is about 3,000,000 miles nearer the Sun in January, when it is closest, than it is in July when it is at its farthest point from the Sun. The difference in distance causes only a very slight change in the warmth of the Sun's rays, much smaller than the familiar seasonal changes.

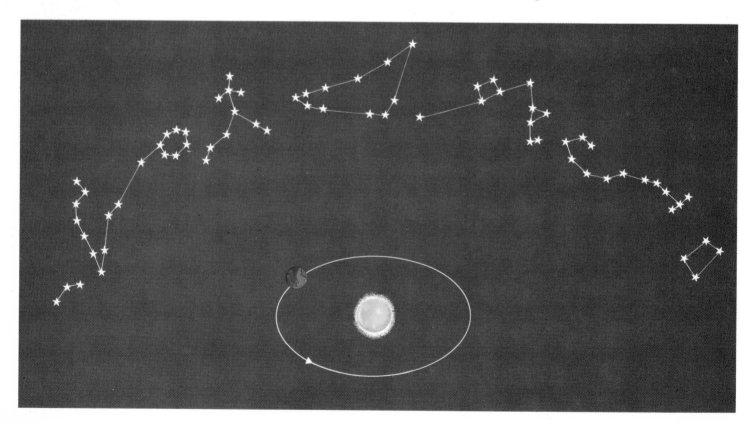

The most important changes during the year for the plants and creatures living on the Earth are the changes of weather that come with the seasons. People, particularly, need to keep track of the seasons so that crops can be planted at the right times. Plants and animals have adapted their behavior to fit in with the seasons. Some animals hibernate during the cold winter. Most plants grow when there is plenty of warm sunlight and remain dormant in winter.

For astronomers, there is another important change during the course of the year. The stars visible at night are different from season to season. Each night, the stars rise above the horizon about four minutes earlier than they did the previous night. After an interval of fifteen days, the stars are rising and setting a whole hour earlier. Over the space of six months, the stars' rising times change by twelve hours. Consequently, the night sky of winter looks completely different from the pattern of stars visible in summer.

Here is an easy way to understand how the seasonal changes in the night sky happen. Think of the Earth at a certain place in its orbit, in December, for example. The night time side of the Earth faces a particular direction in space, so the stars in that part of space are visible at night. After six months have passed, the Earth has traveled to the opposite side of the Sun. The stars that could be seen at night in December are in the same part of the sky as the Sun. They are invisible because they are only above the horizon during the day, when the Sun outshines them. The night time side of the Earth now faces the opposite direction in space. The night sky of July displays a set of stars that are different from those that can be seen in December.

When a whole year has passed, the same familiar stars once again become visible. Although the weather may be quite different from year to year, the changes in the sky are the same. The positions of the stars tell us where the Earth is in its journey and serve as a basis for establishing a calendar.

The drawings at right show the changing winter sky. At 9 p.m. an observer at latitude 50° North will see the constellation of Orion appear in these places on nights in early January (top), February (center) and March (bottom).

The Seasons and the Calendar

As summer changes to fall and winter, the days get shorter and the nights longer. Each day the Sun climbs a little less high in the sky. With the onset of winter in the northern hemisphere, it rises and sets farther south. These changes occur because the Earth's rotation axis is tilted at an angle to its orbit around the Sun. If the Earth's axis were at right angles to its orbit, there would be no seasons. In fact, it is slanted by $23\frac{1}{2}°$ and so the seasonal changes occur.

When summer comes to the northern half of the Earth, the North Pole is tilted toward the Sun. The Sun gets overhead at more northerly latitudes each day until about June 21. On this day, which is properly called the summer solstice, the Sun stands overhead at the Tropic of Cancer, latitude $23\frac{1}{2}°$ North. The summer solstice is the longest day.

Places inside the Arctic Circle (latitude $66\frac{1}{2}°$ to $90°$ North) experience some days when the Sun never sets. These places are tilted toward the Sun even at midnight. Thus countries in the far north became called "the Land of the Midnight Sun." The northernmost city of the United States, Fairbanks, Alaska, lies just south of the Arctic Circle. In the summer it is light enough for outdoor sports at midnight. In winter, however, the long hours of summer daylight have to be paid for. The Sun just manages to crawl $2°$ above the horizon for four or five hours.

While the far north endures the long winter nights, the southern hemisphere enjoys its summer. The Sun has moved southward in the sky. At the fall equinox, about September 23, when day and night are of equal length, it is overhead at the Equator. But later the North Pole tilts away from the Sun, and the South Pole toward it. The winter solstice for the north occurs on the same day as the summer solstice for the south – about December 21. The Sun is overhead at latitude $23\frac{1}{2}°$ South, called the Tropic of Capricorn. When the next equinox comes, on about March 21, the north experiences spring, while fall comes to the south, and the whole cycle starts once again.

The stages of the year are marked off in the calendar, agreed on by everyone, so that appointments can be made and kept, and so that the passage of the seasons can be kept in phase with human needs. The earliest origins of astronomy can be traced to the need to predict seasonal influences on the annual cycle of crop

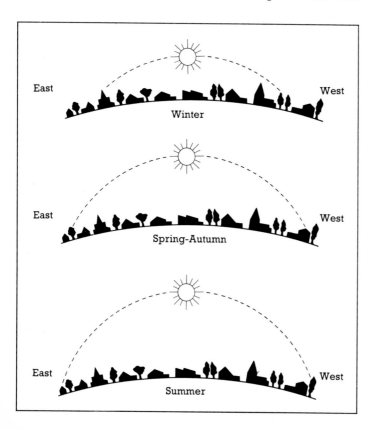

As shown in the illustration at left, the height of the Sun in the sky and its rising and setting points on the horizon change with the seasons. During the winter (top) the Sun is low in the sky and the day is short. In summer, however, the Sun rises and sets much farther around on the horizon and climbs higher in the sky.

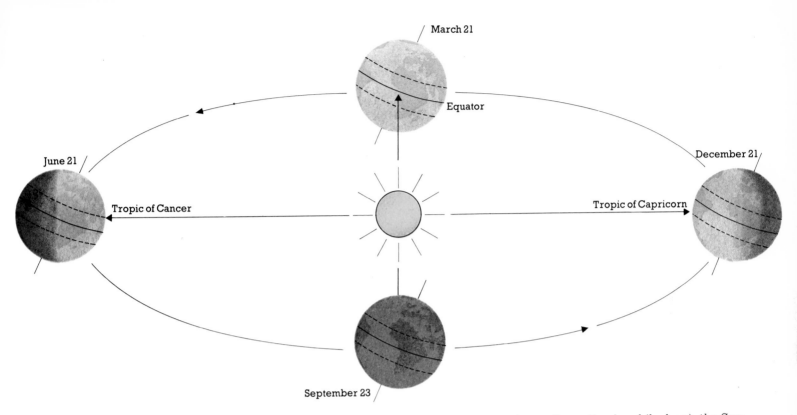

The Earth is held on its elliptical path round the Sun by the gravitational force between the Earth and the Sun. The axis on which it spins is not at 90° to the plane on which the Earth orbits, but is tilted by about 23½° (the amount varies slightly) to make an angle of 66½° with the orbital plane.

At Stonehenge, in southern England (below), the Sun rises above a marker stone outside the main circle of enormous stones. Four thousand years ago or so, Stonehenge was used to find midsummer and midwinter before the invention of writing and the calendar in England.

growth. In the farming communities that existed five thousand or more years ago, a calendar could be established by following the rising and setting points on the horizon of the Sun. This calendar, in turn, allowed the farmers to follow the seasons. In the past great problems arose when the calendar got out of step with the seasons. Observing the stars reveals that a year does not contain an exact number of days. During one circuit of the Sun, the Earth spins on its own axis 365 times – plus a further quarter of a turn. So, there are nearly 365¼ days in a year. To take account of the quarter of a day the calendar year of 365 days adds an extra day in leap years. Leap years are the years that can be divided exactly by four, such as 1980, 1984, etc. However, years ending in 00, like 1900, do not count as leap years unless the year is exactly divisible by 400. An extra day per century must be taken out to remain in step with the seasons.

The Constellations and the Zodiac

For as long as people have observed the night sky, they have imagined the shapes of creatures and objects in the patterns of stars. These star patterns are called constellations. Some well-known constellations are easy to pick out. Others are made up of faint stars that are hard to see except in very dark places away from city lights. Some of the constellation names used now are thousands of years old, while others are comparatively modern. The southern half of the sky has mostly modern constellation names because European astronomers did not begin to catalog the southern hemisphere until the seventeenth century.

The older constellations feature some of the characters from Greek mythology. They include the hunter Orion with his dog Canis Major, Perseus and the Princess Andromeda, whom he rescued from a monster. Andromeda's mother, Cassiopeia, and her father, Cepheus, are there in the sky too. There are many stories to be told in the stars.

For astronomers today the word "constellation" has a special meaning. The sky is divided up into eighty-eight areas, in much the same way that a country is divided up into states or provinces. Each of these areas is called by the popular constellation name for the bright stars within it. All the stars, even the faint ones, inside the boundaries of a constellation belong to it. The official names for the constellations are in Latin and these names are understood by astronomers of all nationalities.

Twelve of the old constellations are well known to most people because they belong to the so-called zodiac. The zodiac is a band of constellations circling the sky. This band is important because the paths of the Sun, the Moon and the planets, except Pluto, are always within it. In a picture of the Earth in its orbit around the Sun, the stars of the zodiac constellations seem to encircle it. The paths of all the planets around the Sun, and the orbit of the Moon around the Earth, all lie very nearly in the same plane. The shape of the solar system is a large, thin disk.

Astrologers are people who study and believe that the exact positions of the Sun, Moon and planets against the background of the zodiac have an important bearing on events on Earth. These positions at the moment of birth form a person's horoscope.

A person's zodiac sign is the one in which the Sun

The twelve signs of the zodiac, devised by the ancient Greeks, are named for a series of constellations. During the year the Sun's path through the heavens passes through each constellation sign of the zodiac in turn.

♈	**Aries**	*March 21–April 20*
♉	**Taurus**	*April 21–May 21*
♊	**Gemini**	*May 22–June 22*
♋	**Cancer**	*June 23–July 23*
♌	**Leo**	*July 24–August 23*
♍	**Virgo**	*August 24–September 23*
♎	**Libra**	*September 24–October 23*
♏	**Scorpius**	*October 24–November 22*
♐	**Sagittarius**	*November 23–December 22*
♑	**Capricornus**	*December 23–January 19*
♒	**Aquarius**	*January 20–February 19*
♓	**Pisces**	*February 20–March 20*

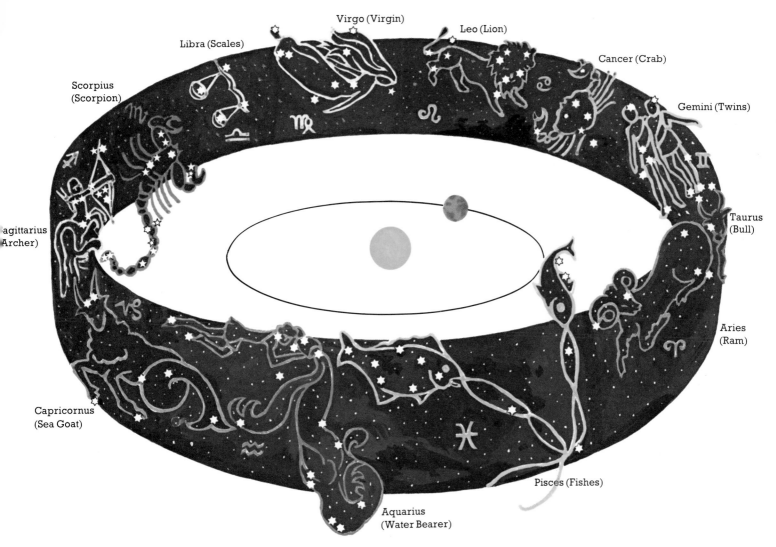

Virgo (Virgin)

Libra (Scales)

Leo (Lion)

Scorpius
(Scorpion)

Cancer (Crab)

Gemini (Twins)

Sagittarius
(Archer)

Taurus
(Bull)

Aries
(Ram)

Capricornus
(Sea Goat)

Pisces (Fishes)

Aquarius
(Water Bearer)

The diagram above illustrates how the twelve constellations in the zodiac encircle the Earth and Sun. In the position shown, the Sun is in Capricornus.

lay at his or her birth. Although the stars cannot be seen during the day, the stars are really still there. As the Earth orbits the Sun in the course of a year, the position of the Sun as viewed from the Earth seems to move through the twelve signs of the zodiac. Astrologers divide the zodiac into twelve equal parts, though the constellations are not all the same size. So, from the astrologers' point of view, the Sun spends just one month in each sign.

From simply looking at the star patterns in the sky, there is no way to tell immediately whether the stars are all at the same distance from the Earth or whether they are scattered through space. But evidence that the stars are at a great distance compared with the Sun, Moon and planets is the fact that their patterns never change as the Earth orbits the Sun.

Ancient astronomers thought that the stars were all fixed on a great, distant sphere. The stars are actually scattered through space at vastly different distances. Modern astronomers have many special ways of finding the distances to the stars. They have instruments capable of detecting very small changes in a star's position. They can measure alterations in the light from stars. Measurements like these allow them to calculate star distances. The star patterns visible from Earth are a result of the planet's position in space. From a planet orbiting another star, the skies would have quite a different appearance.

Sky Maps for Fall and Winter

Just as geographers have made maps of the Earth, astronomers have drawn maps of the sky, showing the locations of stars that can be seen from a certain spot on the Earth at certain times. A traveler needs only one map of a country because the locations of rivers, cities, and seashores never change. The stars, however, change their positions as the year progresses, so a sky-watcher needs a whole set of maps. The six sky maps at right show the stars that are visible from the northern hemisphere on the dates and times indicated. The maps are in pairs, each one covering half of the sky. The maps in the top row show the part of the sky an observer sees when he faces north. The bottom row shows the half of the sky when facing south.

The maps reveal how the sky changes as time passes. Thus an observer who looks north at 1 a.m. on August 1 will see the constellations in the map at top left. Looking north at the same time on October 1, he will see the stars at top center.

The dots indicating the positions of the stars have different sizes corresponding to the brightness of each star. This brightness is called its magnitude, which is measured on a numerical scale. The brightest stars have the smallest magnitude numbers. The brightest star in the sky, Sirius, has a magnitude of — (minus) 1.4. 61 Cygni, one of the faintest visible stars, has a visual magnitude of 5.6. On these maps the largest dots represent the brightest stars from minus first magnitude to plus second magnitude; medium dots are second or third magnitude stars, and small dots stand for stars fainter than third magnitude.

A sky-watcher can locate stars by using the maps on the right on the date and at the time listed below each one. The degree numbers on the sides of the map reveal where the horizon line will be according to the latitude of the observer's location. The dark blue outlines show the position of the Milky Way.

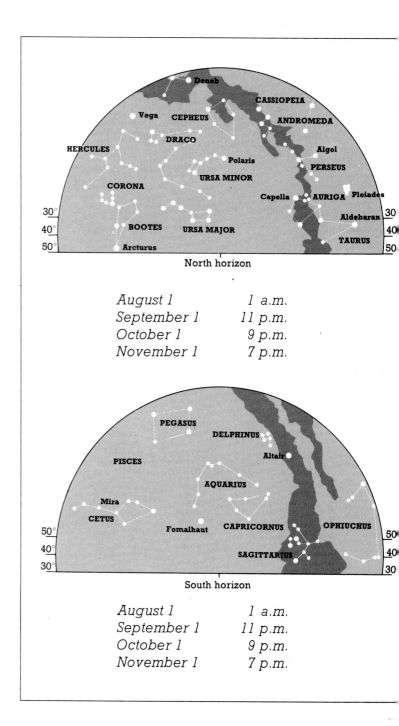

North horizon

August 1	1 a.m.
September 1	11 p.m.
October 1	9 p.m.
November 1	7 p.m.

South horizon

August 1	1 a.m.
September 1	11 p.m.
October 1	9 p.m.
November 1	7 p.m.

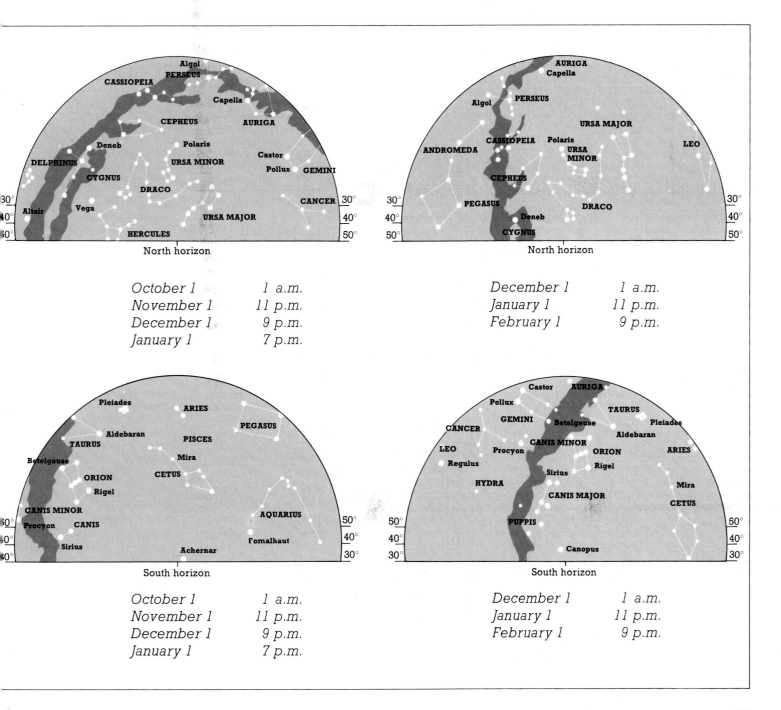

October 1	*1 a.m.*
November 1	*11 p.m.*
December 1	*9 p.m.*
January 1	*7 p.m.*

December 1	*1 a.m.*
January 1	*11 p.m.*
February 1	*9 p.m.*

October 1	*1 a.m.*
November 1	*11 p.m.*
December 1	*9 p.m.*
January 1	*7 p.m.*

December 1	*1 a.m.*
January 1	*11 p.m.*
February 1	*9 p.m.*

Stars: Fall and Winter

A good starting point for recognizing stars in fall is the constellation Cassiopeia. The five main stars of this group make the very obvious shape of the letter W, even though none of them reaches the first magnitude of brightness. This starry W is situated high in the sky during fall evenings. Observers who are located at latitudes around 50° to 60° North see it right overhead.

Two stars in the W of Cassiopeia point to Cassiopeia's husband, Cepheus; and on the other side of Cassiopeia is their daughter, Andromeda. The W is also a signpost to the large constellation of Pegasus, the winged horse. The principal stars in this constellation are part of a large square known as the Square of Pegasus. Looking for this square, bear in mind that it is large and that none of the stars is first magnitude.

Andromeda includes an object of unique interest. The great spiral galaxy M31 is just visible to the eye as a hazy patch of light. Two faint stars in the constellation of Andromeda lead to the galaxy M31, far beyond the edge of the Milky Way. Its light has taken two million years to reach the Earth. The M31 Galaxy is the most distant object visible to unaided human eyes. In a small telescope M31 is a soft glow of light.

These are the six constellations of the zodiac visible in the fall and winter skies in the northern hemisphere.

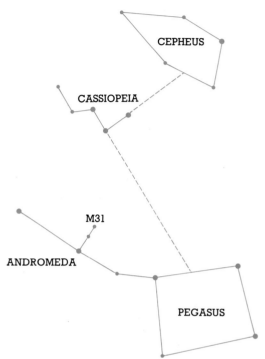

This typical star locator map shows how astronomers use the known position of one constellation, and the stars within it, to locate other heavenly bodies. In the fall sky of the northern hemisphere two "arms" of the W-shaped constellation Cassiopeia point to Cepheus (red line) and Pegasus. M31 can be found by following the map from Pegasus to Andromeda.

Leo   Cancer   Gemini

The bright constellations of winter, when the nights are long, are the best known, as well as being simple to learn.

One of the most splendid views is that to the south on a January evening. Ahead is Orion, marching across the sky, with three stars forming a neat swordbelt for this hunter from mythology. In Orion, seven stars of second magnitude and brighter make up a memorable pattern. Betelgeuse, the star at upper left, is distinctly red. It contrasts well with Rigel, at lower right, which sparkles blue-white. Orion's belt points to twinkling Sirius, also called the Dog Star. Sirius is the brightest star visible from northern latitudes. Procyon is a zero magnitude star, forming a triangle with Betelgeuse and Sirius. Near Orion's belt is a glowing cloud of gas, or nebula, which makes a fine sight through binoculars.

Above Orion are Taurus, the Bull, and Gemini, the Twins. Aldebaran in Taurus is red, like Betelgeuse. Near the Bull's head lies the conspicuous cluster of stars called the Pleiades, or Seven Sisters. Six or seven stars can be seen by the unaided eye, but a small telescope or binoculars will bring dozens more into view. Higher still in the sky lies Auriga, the Charioteer. This constellation contains the yellowish first magnitude star, Capella.

This locator map shows major stars and constellations that can readily be found after Orion has been identified in the winter sky. The dotted lines lead to Procyon and Sirius, to Capella in the constellation Auriga, and to the star Aldebaran and the Pleiades, a star cluster, in Taurus.

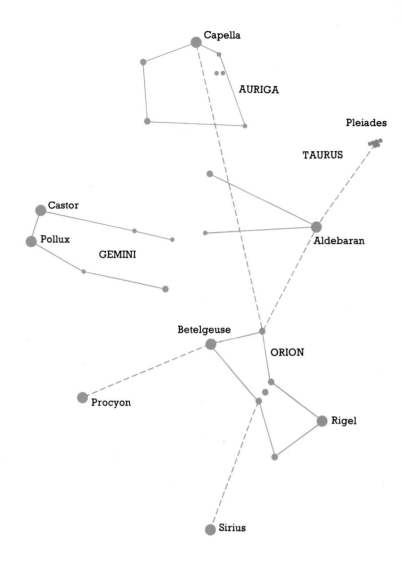

Taurus   Aries   Pisces

Sky Maps for Spring and Summer

At first, in using these maps, the stars directly overhead may be hard to identify. So it is a good idea to begin with the easier constellations. Once these are located, it will be possible to use more detailed maps to identify the fainter stars. Such dim stars are only noticeable to the naked eye when the sky is really dark. The atmosphere picks up any bright light, so the background of the sky tends to be weakly illuminated. In or near a town where there are many street lights the sky never gets really dark, though it may still be possible to identify some of the brighter constellations. If the Moon is up as well, especially if it is nearly full, the sky will be too bright for the observer to see more than a very few stars.

The Milky Way, shown on these maps in darker blue, can best be seen on a dark night, away from street lighting, as a hazy band of light, which is the combined glow of billions of faint stars in our own Galaxy.

After a few minutes in darkness, the eyes will start to adapt. The pupil of the eye opens up to let in more light and is soon able to see fainter stars. Once the eyes have become dark-adapted, it is best to avoid going back into a brightly lit place until observations have finished. For looking at the star maps outdoors use a dim red flashlight. Red light will not affect dark-adaptation very much.

It is a good idea to learn the main constellations one by one. Look for those already recognized and identified on a previous occasion, and try to find a new one, gradually increasing the list.

These pairs of star maps show the major constellations of spring and summer at corresponding dates and hours. Since summer nights are shorter, fewer constellations can be seen than in fall and winter; only three dates and times are given for the maps.

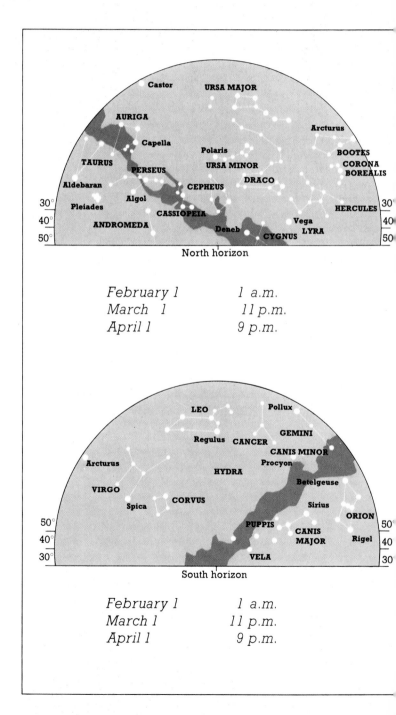

February 1	*1 a.m.*
March 1	*11 p.m.*
April 1	*9 p.m.*

February 1	*1 a.m.*
March 1	*11 p.m.*
April 1	*9 p.m.*

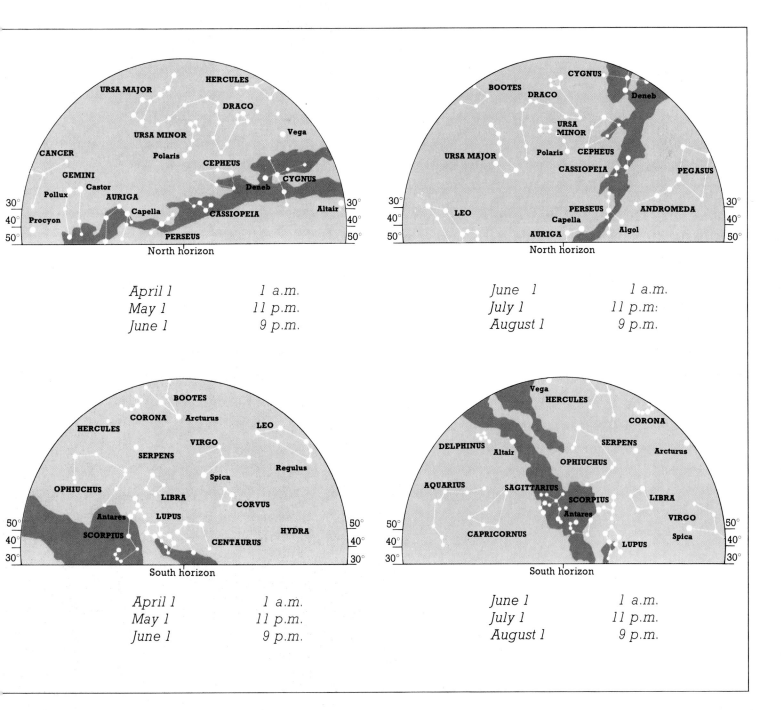

North horizon

April 1   1 a.m.
May 1   11 p.m.
June 1   9 p.m.

North horizon

June 1   1 a.m.
July 1   11 p.m.
August 1   9 p.m.

South horizon

April 1   1 a.m.
May 1   11 p.m.
June 1   9 p.m.

South horizon

June 1   1 a.m.
July 1   11 p.m.
August 1   9 p.m.

Stars: Spring and Summer

In April the constellation Ursa Major reaches its highest point in the sky in the early evening. Observers between latitudes 50° and 60° North, in Canada or Alaska, for example, see it directly overhead. A section of Ursa Major, the Big Dipper, can be used to find three of the first magnitude stars in spring skies in the following manner.

Following the curve of the Big Dipper's tail leads to Arcturus, one of the brightest stars in the northern part of the sky. A reddish star, Arcturus belongs to the constellation of Boötes, the Herdsman. The eye, continuing to sweep in an arc from the Big Dipper through Arcturus, turns the observer's face to the south, and leads to Spica. This white star, the brightest in constellation Virgo, is first magnitude.

Yet another star of the first magnitude, Regulus, can also be seen by facing south. Regulus is the major star in the constellation of Leo, the Lion. It appears in the south in wintertime and remains there through the spring. One way to find this star is to follow a line through the bowl of the Dipper, pointing away from the North Star. Regulus will be the brightest foot – just below the lion's head.

These are the six constellations of the zodiac visible in the spring and summer skies of the northern hemisphere.

Stars in the Big Dipper (within Ursa Major or the Great Bear) can be used in the spring to find Regulus – the brightest star in the constellation of Leo – Arcturus, the brightest star in Boötes, and its neighboring constellation of Corona Borealis, the Northern Crown. Arcturus, in turn, points to Spica, the brightest star in the constellation of Virgo.

URSA MAJOR

BOÖTES

CORONA BOREALIS

Arcturus

LEO

Regulus

VIRGO

Spica

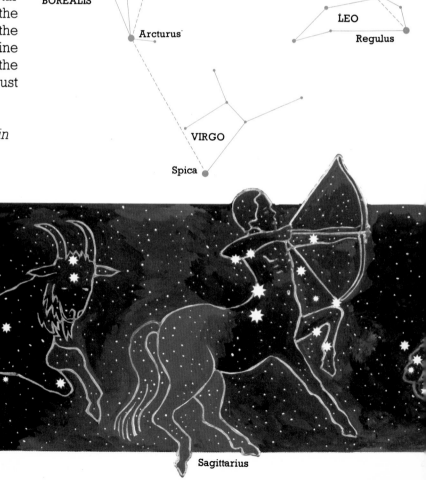

Aquarius

Capricornus

Sagittarius

The locator map right traces a triangle of very bright stars that are visible in the northern hemisphere in summer: Deneb, in the constellation of Cygnus; Vega, in Lyra; and Altair, in Aquila. Stars in the small constellation of Delphinus are not nearly as bright as those of the triangle, but the compact constellation is easy to find near Altair.

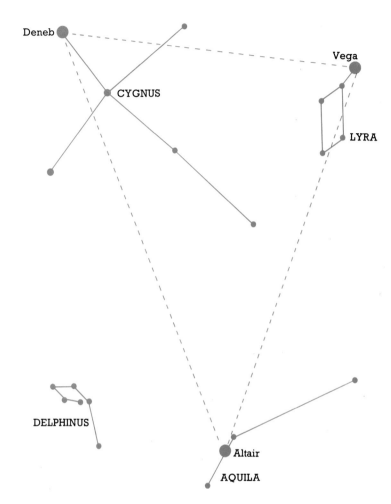

In the skies on summer evenings three stars of first magnitude or brighter stand out: Deneb in Cygnus, Vega in Lyra, and Altair in Aquila. For an observer facing south, these map the Summer Triangle.

Deneb is the brightest star in the constellation of Cygnus, the Swan. This group of stars is sometimes called the Northern Cross. It lies right within the Milky Way. Binoculars or a small telescope can be swept slowly around this part of the sky. Rich star fields, many of them thousands of light years away, will come into view. They make up the soft glow of light from the Milky Way. Close to Cygnus there are dark patches where dust clouds in deep space cut out the faint background of light from distant stars.

Vega is a member of the small constellation of Lyra, the Lyre, and Altair belongs to Aquila, the Eagle. The other stars in these two constellations are much fainter than Vega and Altair. The attractive grouping of stars making up Delphinus, the Dolphin, lies close to Aquila.

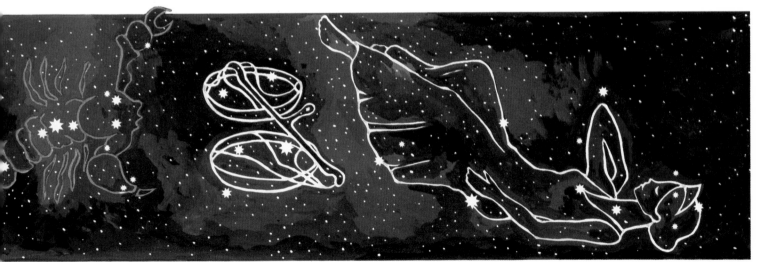

Libra   Virgo

The Sun's Family

The first astronomers, long ago, noticed five special "stars" that gradually moved through the constellations. The Greeks called them *planetoi*, the wanderers, from which came our word planet. Planets shine with a steady light, but real stars often twinkle. This is because a planet is, in fact, a disk of light, whereas a star is so distant that it is always just a point of light. The light from a point source shimmers as it passes through the Earth's atmosphere.

Planets are not like stars at all. The Sun is a typical star. It radiates heat and light of its own, but the planets shine only by the light they reflect from the Sun. Most stars are much larger than planets. The Sun is a thousand times more massive than the biggest planet, Jupiter. The twinkling stars are other suns, much farther away from Earth than any planet.

All the planets visible in the night sky are members of the Sun's family, or solar system. The five planets that can be seen without the aid of a telescope are Mercury, Venus, Mars, Jupiter and Saturn. Mercury is closest to the Sun. It is not easy to pick out because it is never far from the Sun in the sky. Venus is also closer to the Sun than is the Earth. This brilliant planet is seen at its best at dawn or dusk and so it is often called the morning star or evening star. Mars is "the Red Planet," so named because of its color. Jupiter and Saturn, both of them giant planets, can often be seen shining with a steady yellow light.

After the invention of the telescope, astronomers found three, more distant planets. Uranus was discovered in 1781, Neptune in 1846 and Pluto in 1930. All nine planets travel in orbits around the Sun. They all journey in the same direction. The planets closest to the Sun take the least time in orbit. Mercury, nearest to the Sun, makes a circuit in only 88 days, Earth takes a year, and Jupiter almost 12 years.

Studying the motion of the planets, the German astronomer Johannes Kepler discovered in 1609 that the orbits of the planets are slightly stretched circles, called ellipses. An ellipse has two focal points. For each planetary orbit the Sun is at one of the focuses. This means that the distances of the planets from the Sun change slightly as they travel in their orbits.

Kepler found out how the planets move, but it was Isaac Newton, the seventeenth-century English mathematician, who realized that gravitational force holds the planets in their orbits. The Earth's gravity makes objects that are dropped fall to the ground. If the Sun's gravity did not constantly keep tugging at the planets, they would fly off into the depths of space.

The Sun's family has other members apart from planets. Swarming between Mars and Jupiter are thousands of asteroids or minor planets. Comets with their streaming tails approach the Sun from the farthest parts of the solar system. In addition, dust is scattered in the space between the planets, as well as stones called meteoroids. These space rocks burn up if they crash through the Earth's atmosphere, creating a meteor trail, or shooting star.

Many of the planets have moons orbiting them, rather like miniature solar systems. Jupiter has at least sixteen moons, four of which can be seen in a small telescope. Gravitation holds the moons in their orbits around their planets, just as it keeps the whole of the Sun's family together.

The exploration of most of the planets in the solar system is a major scientific achievement of the twentieth century. Men in space have landed on the Moon, and brought back samples from its surface. The five planets that are visible to the naked eye – along with other moons from those of Mars to the satellites of Saturn – have been investigated and photographed by unmanned spacecraft. Other planets are at present the subjects of similar scrutiny; Uranus and Neptune are scheduled for examination by the end of the 1980s, and there are plans to intercept and investigate Halley's Comet with spacecraft in the mid-1980s.

The orbits of the planets are shown in this schematic drawing. The planets are drawn in their correct relative sizes, although they are greatly enlarged compared to their orbits. Between the orbits of Mars and Jupiter lies a broad band of asteroids. All orbits in the planetary system, except those of comets and Pluto, are close to one plane.

The Planets					
	Distance from the Sun				
Name	In millions of miles	Compared to Earth	Time to orbit the Sun in years	Mass compared to the Earth	Radius compared to the Earth
Mercury	36	0.39	0.24	0.06	0.38
Venus	67	0.72	0.62	0.82	0.95
Earth	93	1.00	1.00	1.00	1.00
Mars	142	1.52	1.88	0.11	0.53
Jupiter	484	5.20	11.86	318	11
Saturn	887	9.54	29.46	95	9
Uranus	1783	19.18	84	15	4
Neptune	2794	30.06	165	17	4
Pluto	3706	39.44	248	0.0024	0.28–0.34

This table lists the nine planets in the order of their distances from the Sun. Together with their moons, the asteroids and comets, these planets make up the solar system. The comparisons of the characteristics of the other planets to those of the Earth are given in decimal ratio, that is, Mars is 1.52 times farther from the Sun than is the Earth.

Mercury Venus Earth Mars Moon Jupiter Saturn Uranus Neptune Pluto

Telescopes and Observatories

A telescope's main job is to capture radiation from planets, stars and galaxies. This radiation may be in the form of light waves, radio signals, ultraviolet rays or X-rays. Each type of radiation needs a different type of telescope. Modern astronomers use optical telescopes, infrared and radio telescopes, and ultraviolet, X-ray and gamma ray detectors. The earliest optical telescopes were refracting telescopes. A refracting telescope has a lens to gather light and to form an image of the object. This lens, positioned at the front of the telescope, is called the objective. An eyepiece made from one or more small lenses is used to look at the image made by the objective. The world's largest refracting telescope has a lens 40 inches across. Built in 1897 for the Yerkes Observatory in Wisconsin, it is still in use.

Many problems make it impossible to manufacture lenses any larger than this; for the biggest optical telescopes, astronomers have to use a curved mirror to reflect light, thus forming an image. The largest optical telescope in the world is such a reflecting telescope. It is in the USSR and it has a collecting mirror 236 inches across. When trained on a star, this telescope has the seeing power of a million human eyes. With such power, a big optical telescope can gather detailed information from the radiation of one star. For example, it can build up a photograph, perhaps over many hours, of stars the human eye could never see.

Until 1945 all astronomy was carried out with optical telescopes, but after that date radio astronomy grew rapidly and with the development of rockets and satellites the other wavebands became accessible for astronomical observations. Radio telescopes use huge dishes as much as 1000 feet across to reflect radio signals onto a radio detector. With these telescopes it is possible to measure the strength of radio waves sent out by stars and gain a variety of information about their movements and composition.

The Very Large Array, as this battery of 27 dish telescopes is called, is set up in a Y-formation in the New Mexico desert. Cosmic radio signals picked up by these 85-foot wide aluminum dishes are analyzed by computer. Their arrangement allows them to reinforce each other and measure radio signals with the accuracy and detail of a single-dish telescope about seventeen miles in diameter.

High above the atmosphere's interference X-ray and ultraviolet telescopes orbit the Earth. To describe the view of the sky in X-rays and ultraviolet rays they send radio messages back to the Earth.

Most major optical observatories are now situated in places where the telescopes are least affected by weather. Many are built high in the mountains, above the clouds and away from city lights. In Southern California the famous Mount Wilson and Mount Palomar observatories have become seriously affected by the blaze of lights from the city of Los Angeles. The main center in North America is now at Kitt Peak in Arizona, where eighteen telescopes are in operation. In the nearby city of Tucson the highway lights are shaded against glare.

Inside its aluminum dome, the Anglo-Australian reflecting telescope at Siding Spring Mountain, Australia, is shown tipped over at an extreme angle to reveal its enormous horseshoe mount. Astronomers operate computers that control the telescope; they can see what is happening by closed circuit television. Several similar telescopes with reflecting mirrors about 150 inches across are used in the Americas.

Kitt Peak National Observatory near Tucson, Arizona, is the primary observing site for optical astronomy in North America, and is the largest observatory in the world. The flash of lightning shows the dome of the 156-inch Mayall telescope crowning the summit.

Astronomy in Space

An important advance in astronomy, made possible by space travel, is the sending of television cameras to planets. Men have landed several times on the Moon. Venus, Mars and the Moon have been probed by instruments that landed on their surfaces. Mercury, Jupiter and Saturn have been photographed by spacecraft that passed close by them.

The most spectacular achievement of space astronomy was the USA's Apollo program of manned Moon landings. For the first time, material for scientists to study was brought back from another world, some 240,000 miles away.

It will be many years before people travel to the planets, even the nearer ones. Mars, which depending on relative positions in orbit may be as little as 35 million or as much as 157 million miles from the

Voyager 1, shown here, and its sister spacecraft Voyager 2, have taken revealing and beautiful photographs of the giant planets, Jupiter and Saturn. Voyager 2 is currently on its way to even more remote planets, Uranus and Neptune, which it is scheduled to fly by in 1986 and 1989 respectively.

By the mid-1980s astronomers expect the Space Telescope to be in permanent orbit around the Earth. In Earth orbit there are no clouds or water vapor to disturb the viewing conditions. Because the telescope will be operated in the blackness of space, it will be able to photograph very faint stars and galaxies. The telescope will be controlled by radio signals from a permanent ground station.

Earth, would require an expedition lasting at least a year. Venus is far too hot and its atmosphere too dense and inhospitable for a visit. Instead, these and other planets have been explored by automatic cameras aboard space probes launched from Earth.

In many ways, photographing planets from a space vehicle is similar to a remote television broadcast. A television camera obtains images of a planet. Sometimes the same scene is shot through a series of different-colored filters. The pictures are sent back to Earth as radio signals. The pictures are then reconstructed by computer, and color photographs can be made if at least two color filters were used on board the spacecraft. Hundreds of thousands of photographs of the planets have been made in this way, from Mercury out as far as Saturn, never less than about 744 million miles from the Earth, and of the moons of Mars, Jupiter and Saturn.

Soviet astronomers obtained the first picture from the surface of another planet in 1975 when their unmanned craft Venera 9 landed there and broadcast pictures from Venus back to Earth over a distance of 75 million miles. American scientists then probed the surface of Mars with the landings of two Viking craft in 1976. Eventually, similar craft may settle on moons of the outer solar system, such as those of Jupiter.

Additionally, space telescopes in orbiting satellites make surveys of stars and galaxies that send out invisible ultraviolet light and X-rays. Until instruments sensitive to these rays could be sent up to a height of 90 miles on rockets, astronomers knew little about X-ray stars. But telescopes now circling the Earth permanently have detected many such stars. Some space scientists believe that these instruments have found X-rays coming from the vicinity of black holes, regions of space that because of the enormous pull of gravity within them cannot emit light themselves.

In the 1970s the United States orbiting observatory, Skylab, monitored the Sun with six telescopes. This showed the great advantage of using optical and ultraviolet telescopes in space. Above the atmosphere there is no interference from the weather. In the mid-1980s the space shuttle will be used to place in orbit and service a 94-inch Space Telescope. This instrument will be able to see details that greatly exceed anything that can now be photographed in astronomy.

The shuttle Columbia returning to Earth after its first voyage into space. The further development of space astronomy depends crucially on the success of the shuttle, which will provide a means of constructing and servicing manned and unmanned observatories in space.

The Planet Earth

Blue oceans and white clouds dominate the view of Earth taken from space by Apollo 17 in 1972. Nevertheless, Africa is visible at the top center beneath the clouds.

Man has been able to study the surface of his own planet for as long as the Earth has been inhabited. Yet it is strange to think that before orbiting spacecraft had actually returned color photographs of Earth, nobody had predicted accurately what it would look like from space. Now the Earth can be seen and photographed as a beautiful blue and white planet. From beneath the spiraling patterns of brilliant white clouds the shapes of the continents come into view.

Many factors make the Earth unique in the solar system. It is the only planet with substantial amounts of liquid water. Oceans cover almost three-quarters of the surface. This vast quantity of water is a powerful force of erosion – the wearing away of the Earth's surface. Weather behavior and long-term changes in climate gradually wear down the continental rocks. Mountains are eroded by glaciers, wind and rain. Mighty rivers etch channels through the rocks and lowland plains, carrying sediment away from one place and depositing it in another.

Erosion has given the Earth a quite different appearance from that of the other planets in the inner solar system. For example, there is little evidence now that Earth was once as pitted with meteorite craters as the Moon. But it is hard to imagine that Earth escaped this tremendous bombardment. Erosion by wind and water has helped heal such wounds.

Unlike the other rocky planets, the Earth has inner layers containing tremendous forces that are very active. Volcanoes and earthquakes, for example, permit Earth to let off pressure from friction and heat that build up inside as the great plates of rock comprising the Earth's surface slowly slide about. Earthquakes, sudden, unpredictable and lethal though they may be, teach geologists about the inner structure of the Earth. Vibrations spreading out from an earthquake are measured and analyzed by scientific instruments all over the globe. These vibrations reveal that

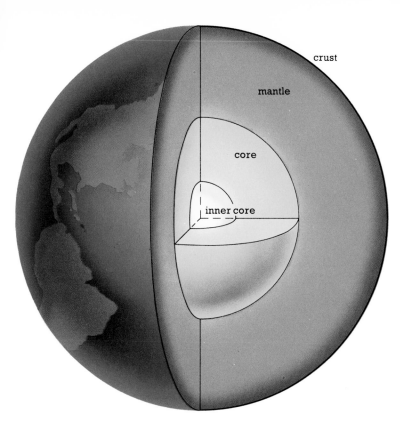

This cutout diagram shows the inner structure of the Earth. The thickness of the Earth's crust may be likened proportionally to the skin of an apple. Beneath it lie the rocks of the mantle and the two-layered core, which is mostly liquid iron.

Earth is made of several layers. On top is a thin crust of rock that is nowhere more than 40 miles thick. The crust lies atop a thick layer of rock 1750 miles deep called the mantle. Inside that, there is a liquid core of hot iron 1400 miles in diameter. Possibly the central part of this core is solid because of the immense pressure created by the weight of the overlying material.

Magnetism is generated by electric currents flowing through the liquid iron in the interior. On most planets a magnetic compass would be of no use for finding north. The compass works here because the Earth has its own magnetic field of influence. A compass needle lines up with the Earth's magnetism and points to the north.

Compared with most of the other rocky worlds in the solar system, Earth is a hive of geological activity. Mountains are constantly being thrust up, earthquakes make the globe tremble and volcanoes cough out liquid rock. Even the continents are slowly gliding about. Only Io, a moon of Jupiter, shows similar activity. Why does the Earth differ from Venus and Mars?

The answer is that the crust of the Earth consists of several large plates that will not keep still. Beneath the oceans and continents there is a rock layer that moves. According to theory, heat flowing from underneath the plates causes this motion, which is like that of a conveyor belt. The heat comes partly from the decay of radioactive rocks. In certain places the plates push into each other, and cause tremendous buckling. This crumpling of two continental plates has caused the formation of the Alps and Himalayas. Along the west coast of North and South America the continental plates are being forced against the oceanic plates and this has formed a great range of coastal mountains from Alaska to southern Chile.

Another effect of these rock movements is to generate friction. This may melt the rock below the surface; molten material works its way upward through cracks and erupts as a volcano.

The motion of continental and oceanic plates is not noticeable in a human lifetime. But it is fast enough to change the face of the Earth. For example, all the present continents resulted when two enormous land masses shattered about 200 million years ago. South America and Africa are still moving about but a look at a map shows how they once fitted together.

Grand Canyon, Arizona, is a dramatic example of the erosive power of water. The Colorado River has cut the face of the desert right down to rock layers two billion years old.

31

Earth and Moon

No small planet has a moon quite like Earth's. Mercury and Venus have none at all, and only two very small chunks of rock are known to hurtle around Mars. In the outer solar system, however, large moons orbit Jupiter and Saturn.

The Earth and its Moon are like a double planet. Mutual gravitational attraction holds them close together. The Moon travels in an orbit around the Earth, taking about a month to do so. (The word "month" comes from "moon".) During the course of a month, the Moon goes through its regular cycle of phases. Just after new Moon, only a thin crescent can be seen close to the setting Sun. In a week the Moon is half lit and after two weeks there is a full Moon. In another two weeks the new Moon reappears. These changes happen because the Moon only shines by reflected sunlight, and as it travels around the Earth, different amounts of the Moon's sunlit half are visible. At its new phase, the Moon lies between the Earth and the Sun, and the side that is shining faces away from the Earth. At full Moon, the Moon is on the opposite side of the Earth from the Sun and the illuminated side is seen.

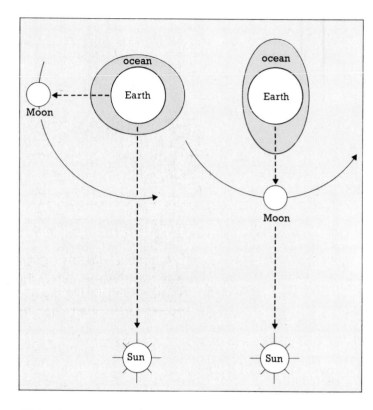

During the course of its monthly cycle of almost 30 days the Moon, seen from the Earth, waxes to fullness and wanes.

This diagram shows how the gravitational pull of the Moon and the lesser pull of the Sun combine to cause the ocean tides.

Waxing crescent   First quarter   Waxing gibbous

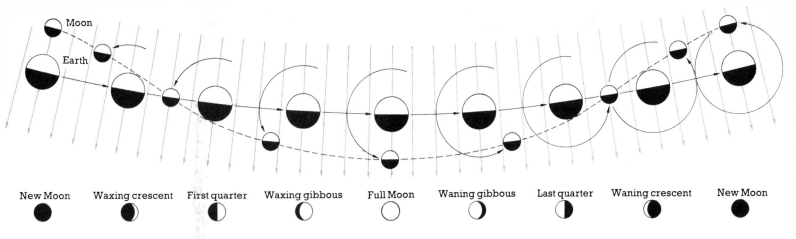

New Moon	Waxing crescent	First quarter	Waxing gibbous	Full Moon	Waning gibbous	Last quarter	Waning crescent	New Moon

Only the side of the Moon facing the Sun is bright. As the Moon travels around the Earth, there is a day – called the day of the new Moon – when no portion of the Moon is visible from Earth. Subsequently, different amounts of the sunlit side can be seen, creating the familiar shapes diagramed in the bottom row above, and shown in the photographs across the bottom of these pages.

Two or three times a year, the full Moon moves into the Earth's shadow. When this happens, the Sun's light cannot reach the Moon because the Earth is in the way, and the Moon is eclipsed. During this so-called lunar eclipse, the shadowed part of the Moon looks dimly red because the Earth's atmosphere scatters reddish sunlight into the Earth's shadow. Eclipses do not take place every month because the Moon's orbit is tilted at an angle to the Earth's path around the Sun.

Usually the Moon clears the Earth's shadow by passing above or below it.

Because the Moon always keeps the same face toward the Earth, the markings on the Moon never seem to change. Until the first spacecraft were sent to travel around the back of the Moon, nobody knew whether it was similar to the visible half. Photographs show that the other side is much the same, except that it is more mountainous.

The Moon's gravitational pull has the important effect of creating ocean tides. The water surrounding the solid Earth is distorted into the shape of a squashed ball under the influence of the Moon's attraction. As the Earth spins on its axis, the bulges in the water seem to sweep around the Earth, causing two tides each day in most places. The Sun, too, influences the tides. When the Moon and the Sun are both pulling from the same direction, the highest tides are formed.

Full Moon	Waning gibbous	Last quarter

33

Eclipses

A total eclipse of the Sun must be one of the most eerily beautiful sights in nature. Only a remarkable coincidence makes it possible for us to witness this spectacle. The Sun is a great, luminous ball, 109 times the diameter of the Earth and at a distance of 93 million miles. The Moon is only one quarter the size of the Earth, but it is 390 times nearer than the Sun and 390 times smaller. Of course, things look much smaller when they are at a great distance than when they are close by. The difference in distance compensates for their difference in size. As a result, the Sun and the Moon appear to be very nearly the same size.

Two or three times a year, on the average, the Moon's path takes it directly between the Earth and the Sun. At such times an eclipse of the Sun takes place. The dark disk of the Moon blots out all or part of the Sun for a short time. Most solar eclipses are not total and even when they are the full spectacle of totality can only be seen over a very small area of the Earth's surface. The Moon's shadow cannot measure much more than 100 miles across on the Earth's surface, and as it speeds along its orbit the shadow sweeps out a long curved path on the Earth.

During a total eclipse of the Sun, an outer crown of light, called the corona, can be seen during the few moments when the disk of the Sun is obscured.

This map shows the predicted tracks of forthcoming solar eclipses during the present century. None will cross North America, though residents of New Guinea will have a chance to witness three.

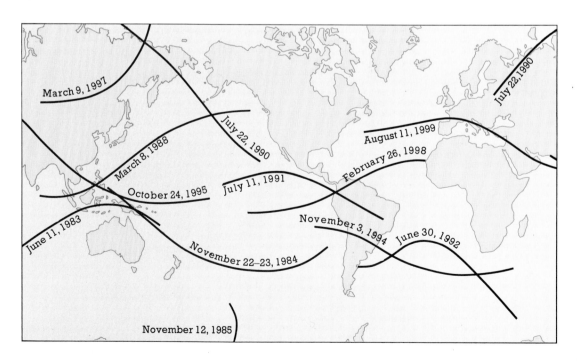

March 9, 1997

July 22, 1990

July 22, 1990

March 8, 1988

October 24, 1995

August 11, 1999

February 26, 1998

July 11, 1991

June 11, 1983

November 3, 1994

June 30, 1992

November 22–23, 1984

November 12, 1985

Any eclipse of the Sun will be seen only by people who are somewhere along this long and relatively narrow path. At any one place the totality lasts for as little as a few seconds to as much as almost eight minutes. An hour or so before totality is due, the Moon starts to cover part of the Sun. At this stage the eclipse is only partial. In a wide area on each side of the eclipse track only a partial eclipse can be seen.

As the total eclipse becomes imminent, the sky darkens and the stars appear. When the Sun's disk is completely blotted out, a shining white halo shimmers around the black Moon. This is the Sun's corona, a crown of thin hot gas streaming away from the Sun. Close to the black disk of the Moon is a thin ring of reddish gas that is called the solar chromosphere. In addition, sometimes visible tongues of gas, called prominences, leap up from the Sun.

The edge of the Moon is not perfectly smooth because there are mountains and valleys running along it. Just before or just after totality the Sun may shine through these valleys. This gives the impression of a string of pearls or a flashing diamond ring.

Just as the Moon causes eclipses, so the Earth itself can cast a shadow on the Moon. This phenomenon, as noted, is known as a lunar eclipse. When the full Moon moves into the Earth's shadow, it can still be seen, shining with a dull coppery light. A lunar eclipse is generally visible across a much wider sweep of the Earth's surface, although eclipses of the Moon are, in fact, rarer than eclipses of the Sun. Up to seven eclipses are seen somewhere on Earth in the course of a year, either five solar and two lunar, or four solar and three lunar. However, the area from which a lunar eclipse can be seen is so much greater in extent than the solar path that, at a given location, more lunar eclipses will be seen over a period of time.

This composite illustration shows the progress of a so-called annular eclipse in Costa Rica over a period of about one hour in 1973. When the Sun is closer to the Earth than usual, and the Moon farther away than usual their sizes do not match, and the Moon fails to cut out the Sun completely.

Because the Earth and the Moon have elliptical orbits, the distances between the Sun and the Earth, and between the Earth and the Moon are not constant. When the Sun is closer than usual and the Moon is more distant than usual, the Moon looks a little smaller than the Sun. If an eclipse then occurs, the Moon does not cover the Sun completely. Instead a bright ring of sunlight circles the black Moon. Such an event is called an annular eclipse; the Latin word *annulus* means ring. During an annular eclipse, the sky remains bright and the corona cannot be seen. For those reasons, annular eclipses are of little scientific value, since they do not allow any unusual observations of either body.

Exploring the Moon

In 1959 a Soviet spacecraft made the first flight beyond the Moon and sent back blurred pictures of the far side. Since then, there have been about forty successful unmanned missions by American and Soviet craft to the Moon. On some of the early attempts the craft simply coasted by the Moon, photographing it as they passed nearby. Later, man-made satellites were placed in orbit around the Moon to make very detailed photographic surveys as American scientists looked for suitable locations for manned landings. The next stage was to test the lunar surface by softlanding unmanned probes. Then, in 1969, the Apollo 11 mission achieved the first manned landing. In all, six Apollo crews visited the Moon between 1969 and 1972.

A proud achievement of the manned space program has been the return to Earth of over 2000 separate samples of Moon rocks, totaling 174 pounds in weight. These samples are kept in a special laboratory in Houston, Texas, where they are protected from contamination by contact with anything on Earth, including the air. The study of this lunar material, offering clues to the Moon's history, will continue for many years.

As a simple example of a scientific problem that was solved by examining this material, we can consider the origin of the Moon. Once it was seriously believed that the Earth and Moon had been a single planet that split in two. The lunar samples showed that the Moon had never been attached to the Earth.

The actual landings required three men in space

Viewed from above the barren surface of the Moon, the planet Earth hangs motionless in the blackness of space. From any given location on the Moon, the Earth always appears in much the same position because the Moon completes a rotation on its axis in the same amount of time that it takes to complete a revolution around the Earth.

on each mission. Once the spacecraft was close enough to the Moon, its rocket engines were used to put the craft into lunar orbit. Then the craft separated into two parts. Two astronauts piloted a Lunar Module (LM) to the surface, while a third colleague kept watch from the orbiting Command and Service Module (CSM). The Lunar Module contained everything needed to survive on the Moon: oxygen, food and equipment. Some of the Apollo missions carried a vehicle called a Lunar Rover for taking drives into the "countryside."

When each exploration was completed, the Lunar Module blasted away from the Moon to dock with the Command and Service Module. All the astronauts then made the return journey to Earth in the CSM, after jettisoning the LM to crash back on the Moon's surface. Just before re-entry into the Earth's atmosphere, the Command Module, with the astronauts inside, separated from the Service Module. The crew maneuvered their capsule so that its heat shield would protect them from intense frictional heating as they returned through the air, until it splashed down in the ocean.

Soviet research has stressed the need to send unmanned robot space vehicles to explore the Moon. By this means Soviet scientists have collected 11 ounces of Moon rock. Their two Lunokhod vehicles were able to trundle about for some months, making a thorough exploration of the regions where they had landed.

Man has explored only a tiny part of the Moon. The space shuttle, America's reusable space vehicle, may allow more trips in the future. One day, perhaps, there may be scientific bases on the Moon, similar to those in the Antarctic today, supplied by regular shuttle deliveries of food or equipment.

In front of the Lunar Module of the Apollo 17 mission in 1972 is the Lunar Rover, a vehicle and mobile TV station, that aided the exploration.

The Command and Service Module for the Apollo 16 mission in 1972 glides in orbit awaiting the return of the manned Lunar Module from the Moon's surface.

Lunar Craters and Seas

A look at the Moon through binoculars reveals that much of its surface is scarred by numerous craters. The darker regions are smoother than the bright parts, with fewer craters. The dark areas were named maria (the plural of the Latin word *mare*, meaning sea) before anything was known about their nature. Less than a century ago, people still thought they might be oceans of water. In fact, billions of years ago, the maria were liquid, but made of molten rock that flooded out over parts of the Moon. This lava turned solid to make the mare plains we see today.

Manned spaceflight has settled the arguments about the origin of the craters. Most of them are the result of intense battering by the interplanetary rocks known as meteoroids, but a few are the remains of burned-out volcanoes. Soon after our Moon and the planets first formed, the solar system was still strewn with stray rocks and boulders. These bodies then pounded the surfaces of the planets and the Moon, briefly melting the rock at the points of impact. The heavy bombardment tapered off about four billion years ago, but cratering has continued until the present day because small meteoroids keep hitting the Moon. The youngest craters on the Moon, such as Tycho and Copernicus, have bright systems of rays fanning out from them. Best seen at full Moon, these rays are caused by fragments of rock that were blasted out of the craters. Other craters have small central peaks or mountains.

What would happen if a large asteroid, weighing many thousands of tons, slammed into the Moon? The remains of such drastic collisions are clearly visible where they dug out deep depressions in the surface long ago. These basins became maria when they were flooded by molten rock welling up from inside the Moon, between three and four billion years ago. The Moon's far side has no large mare. It is nearly all highlands, devastated over the ages by meteoric bombardment.

Evidence of the Moon's fiery, volcanic past can still be seen on the maria, in the form of many deep, winding valleys called rilles. These may be the remains of the ancient channels through which the molten lava flowed. Some of the smaller craters are certainly the cindery holes of dead volcanoes. The maria also have bumps and domes where the pressure of molten rock has lifted up the surface a little. Even today there may be slight volcanic action: observers sometimes notice temporary red hazes along the edges of maria, which could be caused by gas puffing out of weak spots in the lunar crust. Amateur astronomers have made valuable contributions by recording these transient phenomena.

A composite picture of the Moon, made from photographs of two half-moons, clearly shows the lunar craters and, in some places, ray systems of rock thrown out when the craters were formed. The principal seas and craters on the photo are:

(1) Mare Crisium
(2) Mare Foecunditatis
(3) Mare Nectaris
(4) Mare Tranquillitatis
(5) Mare Serenitatis
(6) Mare Frigoris
(7) Mare Imbrium
(8) Eratosthenes
(9) Copernicus
(10) Kepler
(11) Oceanus Procellarum
(12) Mare Nubium
(13) Mare Humorum
(14) Tycho

The Surface of the Moon

The Moon's surface is blanketed by very fine powder. Walking on the Moon is a bit like hiking through talcum powder with stones in it. Boots leave a crisp impression, as in freshly fallen snow, that will last for millions of years.

Moon soil is not at all like Earth soil. It is made entirely from finely pulverized rock – the dust from meteoroid crashes. Moon soil has no water, decaying plant material or life. But it does contain something beautiful and unusual. Moon soil has many glass beads, emerald green and orange-red in color, shaped like jewels and teardrops. These are made when a meteoroid impact sprays liquid rock in every direction. When the droplets of rock solidify, they turn glassy.

On the surface of the Moon a man weighs only one-sixth of his Earth weight. This is because the Moon's mass is a mere one-eightieth of the Earth's, so the gravitational pull is considerably smaller.

It was once feared that if a spacecraft landed on the Moon it would rapidly sink without trace into the deep dust layers. However, the lunar soil – as well as the gravel and stones it includes – is well packed down to provide a reasonably firm surface. The main hazard of Moon travel is finding a smooth place to land. At close quarters the surface looks much like a bomb site, with small craters everywhere.

Moon rocks are distinctly different from Earth rocks. A geologist could easily tell them apart. The difference between them suggests that the Moon was once hotter than the Earth has ever been, and emphasizes the fact that the Moon has no air and no water. The oldest rocks found on the Moon are 4.6 billion years old. In comparison, the most aged rock yet discovered on Earth

The Moon crater Eratosthenes, seen from a spacecraft, has walls rising from above the surrounding plain, but inside the crater they have slipped down to form terraces.

Astronaut Neil Armstrong, the first man to walk on the lunar surface, makes a crisp footprint during the Apollo 11 Moon landing in 1969. It will still be as sharp as this millions of years in the future.

dates from only 3.8 billion years ago.

Astronauts left scientific apparatus on the Moon, including sensors that have detected numerous "moonquakes" as well as the impacts of meteoroids, some spacecraft and man-made debris slamming into the surface. Several small reflectors, like those on a car or bicycle, were placed on the Moon. Scientists can now measure the Moon's distance to within an inch or so by aiming a powerful laser beam at these reflectors and timing the beam's round trip from Earth to Moon and back again. This distance on the average is 240,000 miles.

The exploration of the Moon, and the return of lunar material to the Earth for laboratory analysis, led to entirely new discoveries and areas of research. Geological maps of most of the surface are now available, a possibility undreamed of before about 1960. Samples, mainly from the Apollo program, have been sent to laboratories throughout the world for very detailed examination. Nevertheless, this analysis of lunar material has shown that the surface has never supported life in the past. However, astronauts brought back to the Earth a piece of the Moon lander, Surveyor 3, which had landed on the Moon three years previously. Bacteria on this craft were still alive after several years of exposure to the harsh lunar environment. These bacteria did not flourish, but neither did they die. Thus there is a faint, extremely remote possibility that spacecraft are contaminating the Moon, planets and deep space with microscopic life from Earth even though the equipment is given a complete cleaning before its launch.

Shown above right, the crater Tsiolkovsky, on the far side of the Moon, has a black carpet of lava and a central mountain. This is one of the most beautiful lunar craters.

This ancient crater, called Thomson, has been flooded with lava, which has in turn been cratered by minor impacts.

41

Mercury

Mercury is a planet of extremes. The closest planet to the Sun, it has scorching daytime temperatures of higher than 800°F, hot enough to melt lead. At night the temperature plunges down as low as −280°F because there is no blanket of atmosphere to trap the heat. Mercury spins around on its own axis in 59 days and takes 88 days to make one circuit of the Sun. The orbit is a pronounced ellipse, so that the distance from the Sun varies between 29 million and 43 million miles. This tiny planet is not much bigger than our Moon.

In 1974, the space probe Mariner 10 flew past Mercury. It sent back to Earth photographs of about one-third of the planet's surface. These were enough to show that Mercury is similar in many ways to our Moon. There are a great many craters and several

This illustration shows the relative sizes of Mars, Mercury, Callisto (a moon of Jupiter) and the Earth's Moon.

smooth plains. Some of the large craters have bright rays spreading out just like the large lunar craters.

Geologic mapping has indicated that Mercury went through a period of intense volcanic activity early in its history. Approximately one-quarter of the surface consists of smooth lava plains.

There is no weather on Mercury, just long baking days followed by intensely cold nights. The surface is, therefore, not changed by erosion. Over billions of years space debris has smashed onto the surface of the planet. This continuous bombardment has left the shattered surface we see today. The crater floors are covered in fine dusty powder, the product of smashed rock.

The heaviest bombardment of Mercury took place long ago, only a few hundred million years after the planet formed. Some of the material smashing into the surface scooped out craters hundreds of miles across. At the same time, volcanoes sprang into action, flooding craters with molten rock. Today, however, the planet Mercury does not seem to have any active volcanoes.

An extremely thin atmosphere of helium gas was discovered by Mariner 10. There is so little of it that the surface pressure is billions of times smaller than at the surface of the Earth. A small planet like Mercury cannot retain this helium gas for long, because it simply evaporates into space. Probably it is replenished from helium that was ejected from the Sun and captured by Mercury. Life like that on Earth could not exist on this airless world with its great range of temperature.

Mercury is a difficult planet to observe. Its close orbit around the Sun means that it is never visible more than 27 degrees in the sky from the Sun's disk. Consequently, Mercury can be seen only just before sunrise or after sunset, close to the horizon. Mercury's rapid motion around the Sun limits the possibility of seeing it to just a few days per orbit. In a telescope it shows phases like the Moon, because parts of the planet are not in sunlight. In searching for Mercury with a telescope, be sure not to point the instrument at the Sun.

The large, fresh impact crater above, which has a diameter of 75 miles, was photographed by Mariner 10 from a distance of 21,100 miles as the spacecraft approached Mercury for the first time on March 29, 1974.

The cratered surface of Mercury shown at left in a composite photograph is similar to the cratered highlands of the Moon. The largest craters visible in this photo are about 125 miles in diameter. Most of the pictures that make up this photomosaic were taken by Mariner 10 from a distance of 125,000 miles.

Venus

Brilliant Venus is one of the easiest planets to recognize in the sky. It gets nearer to Earth than any other planet. At its closest approach, Venus is only 26 million miles away.

The size and mass of Venus are both similar to Earth's, but its atmosphere is very different. The main gases in the Earth's air are oxygen and nitrogen. Instead of these gases, Venus has a suffocating atmosphere of carbon dioxide; and high up there are even misty clouds of sulfuric acid.

One remarkable property of Venus's atmosphere is the way in which it acts like a greenhouse. The glass in a garden greenhouse lets high-energy rays from the Sun pass through to heat the soil. But the glass will not let out the low-energy heat that is produced by plants and soil. So heat is trapped inside the house and the temperature goes up. The dense carbon dioxide gas on Venus works rather like greenhouse glass. It holds in the heat so that the temperature on the surface is nearly 900°F, which is even hotter than on Mercury. There is so much carbon dioxide that it crushes down

The Soviet spacecraft Venera 14 sent back to Earth this panorama of the rock-covered surface of Venus on March 5, 1982. A soil-sampling device is seen in the center of the photograph. Soviet astronomers have had four craft land on Venus, but none survived longer than one hour on the surface of this harsh planet.

with an enormous pressure – almost one hundred times greater than that of Earth's atmosphere. On Venus the weight of air is as great as the pressure one mile beneath Earth's oceans.

Until the space age very little was known about Venus. Swirling clouds hide the surface from view. The blistering heat and crushing atmosphere destroyed many spacecraft that tried to penetrate to the surface. In 1975 two Soviet craft survived long enough to send back the first photographs ever taken of another planet's surface.

In 1978 the two American Pioneer Venus probes reached their target. One was a multiprobe, which released five separate probes at different points in the atmosphere. Each gathered data and relayed its findings back to Earth as it fell through Venus's atmosphere and eventually crashed on the surface.

The other Pioneer Venus craft was an orbiter. One of its main achievements was to map 93 per cent of the planet's invisible surface by means of radar. The radar results are not as detailed as photography would be, but plains, mountains and rolling uplands have been located; and there are circular features that may be craters. Radar has also shown that Venus spins slowly – rotating in the opposite direction to the Earth. It takes 225 Earth days for Venus to orbit the Sun. One Venus "day" lasts 118 Earth days, so there are less than two Venus days in a Venus year.

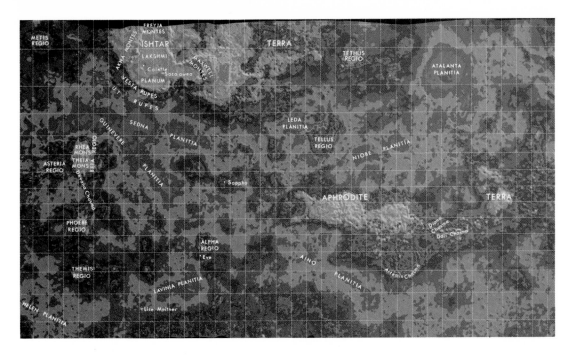

A radar-map, generated by computer, uses colored contours to depict surface features on the planet Venus. Features are named, by international agreement, for female figures in classical mythology and science. The radar data was obtained from the Pioneer Venus I spacecraft orbiting the planet in 1978 and 1979.

A picture taken in ultra-violet light (left) reveals cloud patterns on Venus that are not visible in ordinary light. The top layer of clouds shown here – Venus has three layers of clouds – moves at the rapid rate of 225 miles per hour. The causes of the V-shaped break in the clouds are unknown.

Mars, the Red Planet

Viewed through a telescope, the planet Mars looks like a rusty-red disk. Its surface has various light and dark parts, as well as white ice caps at the north and south poles. At times Mars is sufficiently close to the Earth for it to shine more brightly than any of the stars for a period of several months.

Like Earth, Mars goes through a cycle of seasons – while one half of the planet has summer, the other half has winter. The cold winter causes carbon dioxide gas in the atmosphere to freeze out as a polar ice cap. In the summer half of the planet, the polar cap gets smaller because of the Sun's warmth. Meanwhile, the other half of Mars is having winter, and, as the ice forms again, its polar cap grows.

During the course of a Martian year, which is nearly twice as long as an Earth year, changes occur in the appearance of the surface, especially near the ice caps. The markings on Mars and their changes led astronomers to speculate for many years that Mars might have simple plant life. Earthbound telescopes could never answer questions about life on Mars for

These three photographs show changes in the appearance of Mars. The white polar ice cap is largest at the left (August 21, 1971). The center view was taken on October 9, 1973, and only six days later (right) the planet was partly obscured by a dust storm. These photographs were taken at Lowell Observatory, Arizona.

certain, but now Mars has been visited by spacecraft that have actually landed on it.

Viking 1 and Viking 2 both landed on Mars in 1976. They sent back to Earth many marvelous color photographs. The Viking craft also made experiments to find out about the soil and the atmosphere. Apart from these landers, several spacecraft have orbited around Mars. They have taken thousands of photographs from space so that a very great deal is now known about what Mars is really like.

The whole planet is a great desert. No water flows on the surface and practically none exists in the atmosphere either. Color photographs show a completely barren landscape, strewn with loose boulders. The red color is typical of desert rocks that are found in many places on Earth, and comes from the oxidation or rusting of iron. Only the surface is this color; the Viking landers scooped up the red dust and found that just below the surface the rocks are a darker color.

Even the sky looks red on Mars, due to red dust in the air. Sometimes great dust storms develop, and about every two years there is such a huge hurricane that the whole planet becomes engulfed in choking dust. Currents of air and very strong winds blowing at hundreds of miles an hour whip up the dusty surface into billowing clouds. When the storm dies and the dust eventually settles, it drifts to form lighter colored layers in some places. In other places, the dust is swept away to show the darker rock underneath.

A model of the Viking lander (left) extends its sampler arm, devised to bring scoops of Martian soil into the lander for analysis. The two cylinders with slits are cameras for scanning the planet. These devices, as well as a seismograph, a weather monitor and other sensors, can radio their findings back to Earth through the dish antenna on the top of the lander.

Viking 2 took this picture of the Martian surface three years after its landing. The photo shows part of the spacecraft itself (right), several small trenches where the sampler arm scratched up material for analysis, and the arm's protective shield (on the ground), which was shed after the landing. Most rocks in the photo are one to two feet wide. This is one of the 4500 photos Viking 2 transmitted from Mars.

The Surface of Mars

Like the Moon, Mars is pitted by many craters. These were created by meteoroids that crashed onto the surface from space. Volcanic activity has also contributed to the scenery on Mars. In the region called the Tharsis Ridge there are three gigantic extinct volcanoes. These mountains are called Ascraeus, Pavonis and Arsia. Photographs taken from orbiting spacecraft show many old lava flows, long since cold and solid. The Viking landers took pictures of rocks that have a bubbly structure. This type of rock is made from volcanic lava. When the lava becomes solid, gas bubbles leave small holes in the rock.

Another dramatic feature is the Mariner Valley, stretching nearly one-fourth of the way around the planet. It makes the Grand Canyon look very small, being three or four times as deep, and long enough to stretch about halfway across the United States.

Mars is an inhospitable place. This cold, dry world has an atmosphere that is very thin compared with Earth's air. The Martian air consists mostly of carbon dioxide and has so little oxygen that people and animals would not survive breathing it.

No vegetation grows on Mars, not even simple plants such as moss and lichen. However, if there is any life at all, it must be in the form of a lowly plant life that has escaped detection. Although there is probably no life whatsoever now, it may have been very different in the past. Dried-up riverbeds and places where water has flooded over the surface suggest that Mars once had plenty of water flowing in streams and rivers. However, it must be a billion years or more since the last rain fell on Mars. One of Mars' great mysteries is the present location of this water, which is definitely no longer on the planet's surface. Small, thin clouds can be seen on many of the Mars orbiter photographs, but these cannot make any rain.

Billions of years ago the Martian atmosphere was much thicker, like Earth's. However, Mars is small, only just over half the size of the Earth. Its gravitational pull is less than Earth's and so the atmosphere has gradually drifted off into space.

Today the only water on Mars appears to be in the parts of the northern polar ice cap that never melts. The frost that comes and goes each year at both poles is actually frozen carbon dioxide, sometimes called dry ice. Close-up photographs of the polar caps show a surprising effect. All around the pole, dark bands curve in spiraling patterns. These bands are valleys that are free of frost. Inside the valleys are rocky terraces revealing many layers of mixed snow and dust. The winds of Mars circulating around the polar caps have caused the valleys to be eroded away, and ridges left standing. Around the north polar region there is a giant ring of sand dunes. The red sands are constantly shifting to and fro over the uninhabited wastelands of this dry, desert world.

The relatively thin atmosphere of Mars causes it to have high wind speeds, of several hundred miles an hour. These winds stir up immense dust storms, particularly when the planet is at the part of its orbit closest to the Sun.

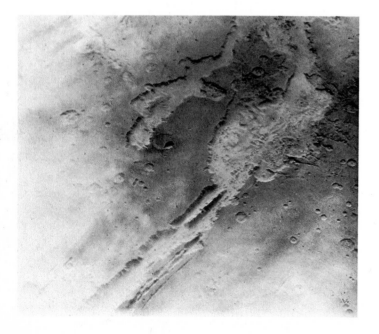

Viking 1 photographed part of the Mariner Valley, a vast chasm on Mars that is in places 150 miles wide and several miles deep. Rivers, now dry, have cut deep channels in the side of this valley.

About 4 miles high and 200 miles across, the volcano in the photo at left is one of the smaller ones on Mars. The numerous craters on the volcano suggest that it is also among the oldest on the planet.

Arsia Mons (below) is one of the enormous dormant volcanoes in the Tharsis region near the equator of Mars. The mouth of the volcano is nearly 90 miles wide and the mountain below it reaches to 87,000 feet, nearly three times the height of Earth's Mt Everest.

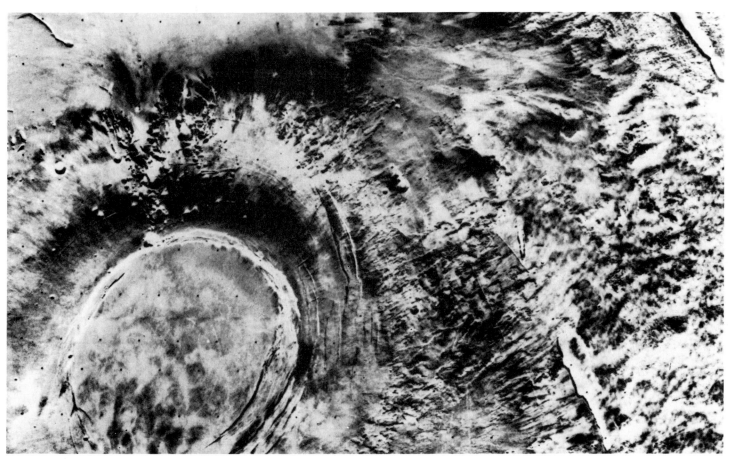

Asteroids and Meteors

In 1801 the Italian astronomer Giuseppe Piazzi discovered a new object between Mars and Jupiter, which he named Ceres. Since it measured only 250 miles across, it did not qualify as a real planet. It proved to be the first of many similar minor planets, or asteroids, found by later astronomers, most of them cruising around the Sun between the orbits of Mars and Jupiter.

Today astronomers keep track of approximately 3000 asteroids, and new ones are continually being found. They are irregular lumps of rock, and most of them measure only a few miles from end to end. The two midget moons of Mars, named Phobos and Deimos, are similar to asteroids, and close-up photographs show that even these small objects have a sprinkling of craters. At one time astronomers theorized that the asteroids were the wreckage of a planet. However, the total amount of matter in the asteroid belt is less than one-hundredth the mass of the Earth; so this theory is unlikely. Very probably, the asteroids are the remains of material left over when the planets formed about five billion years ago.

Phobos, one of two tiny moons of Mars, is perhaps very similar to a small asteroid. This satellite is roughly ellipsoidal in shape, with dimensions of 12 x 13 x 17 miles.

When the Earth passes through a stream of meteoroids, the bright trails appear to fan out from one part of the sky. This time exposure, taken during the Leonid shower, captures the tracks of several meteors. The star images are drawn out into short lines because of the Earth's rotation during the course of the time exposure.

Interplanetary space seems to be full of small bits and pieces of rock and dust, called meteoroids. Every day the Earth collides with millions of these objects. Fortunately, the atmosphere forms an effective shield against meteoroids. Most of them burn up by frictional heating in the atmosphere before reaching the ground. Some 1000 tons of space dust from these burned-up rocks fall to Earth every day. Some meteoroids, when they heat up, make trails bright enough to be seen by the naked eye at night as they whiz through the air. Such momentary streaks of light are "shooting stars," but astronomers describe them as meteors. An exceptionally bright meteor is known as a fireball.

The meteoroids in space orbit the Sun, just like the planets. Sometimes the Earth brushes through a whole swarm of meteoroids. Then there may be a wonderful display of meteors, perhaps one a minute. In such a meteor shower, the bright streaks radiate out from one part of the sky, called the radiant. A shower is named after the constellation in which its radiant lies. Some meteor showers occur regularly, every year. The Leonids, for example, with the radiant in the constellation Leo, appear in November. In 1966 the Leonids gave a most spectacular display, sixty thousand meteors passing overhead within 40 minutes. A spectacle on this scale is, however, a fairly rare occurrence.

A rocky object from space that actually crashes right onto the surface of the Earth is called a meteorite. A meteorite is normally larger than a meteoroid, often weighing several pounds. Fragments of meteorites are eagerly examined by scientists, who are interested to find out what they are made of. Most are stony, but about five per cent are composed of iron and nickel. A very few meteorites have traces of organic material, such as amino acids, one of the building blocks of life. This demonstrates that the simplest substances associated with life exist naturally in astronomical bodies.

Sometimes very large meteoroids crash onto the Earth, with spectacular results. About 30,000 years ago an iron meteoroid, 50–100 feet across, hit the Arizona desert. It made impact at 10 miles per second. A tremendous explosion ripped a crater 565 feet deep and three-quarters of a mile wide. This crater is now a major tourist attraction. Fortunately, such large meteoroids are rare.

A lump of iron as large as an apartment house smashed into the Arizona desert 30,000 years ago and left this crater near the present site of the town of Winslow. The Earth has few major meteor craters; most have long been erased by weather and water.

This large iron meteoroid fell to the ground thousands of years ago near Grand Canyon, Arizona. It is chained to the ground, but this precaution scarcely seems necessary since the meteoroid weighs 535 pounds.

Jupiter

Jupiter, chief of the ancient Roman gods, is a fitting name for the giant planet of our solar system. At times Jupiter outshines all the stars in the night sky – only the planet Venus gets brighter. Jupiter's brilliance results from its immense size; its diameter is eleven times that of the Earth, and its volume over one thousand times as great. Indeed, there is more matter contained in this giant than in all the rest of the planetary system put together.

A small telescope shows up the bands of colored clouds that cross Jupiter's disk. Astronomers monitor the fluctuations in color and position of these clouds as well as they can. Because Jupiter whirls around on its rotation axis in less than ten hours, astronomers can view much of its visible clouds in a single night. But the most detailed and magnificent observations of the planet have been obtained by space probes sent to explore the outer reaches of the solar system.

Pioneers 10 and 11 were the first missions to Jupiter. In 1973 and 1974 they returned the best pictures of the planet ever seen up to that time. The two Voyager craft that reached Jupiter in 1979 provided an incredible wealth of images of the planet and its family of moons, plus other scientific data.

Because of Jupiter's great mass, its gravitational pull, compared with that of the Earth, is very strong. As a consequence, Jupiter still contains most of the material it had when originally formed, because even

In this photo of Jupiter, taken from a range of 20 million miles, cloud features only a few hundred miles across are visible. In the southern hemisphere the atmospheric circulation is dominated by turbulent cloud and the magnificent Great Red Spot.

the lightest gases cannot escape its gravity. The chemical composition of Jupiter is therefore very similar to the Sun's: primarily hydrogen, with about one-tenth helium and much smaller quantities of the other elements. The inside of Jupiter is very hot, possibly about 50,000°F. Scientists believe that Jupiter is mostly fluid – largely liquid hydrogen – although the core may be solid rock.

Electric currents flowing in the interior of Jupiter create the strong magnetic field that surrounds the planet. This is four thousand times stronger than the Earth's magnetism.

The swirling patterns of clouds and spots that astronomers see from Earth are in Jupiter's upper atmosphere. The rapid rotation, the heat welling up from inside, and the sunlight falling on the topmost layers, all contribute to the driving force of Jupiter's "weather." A remarkable feature, oval in shape, which has intrigued astronomers for three centuries,

is the Great Red Spot. This and other visible ovals seem to be anticyclones, which circulate winds around regions of high pressure. The Great Red Spot has been steadily visible for at least 140 years. Its color, which may arise from the presence of such chemicals as phosphorus, sometimes fades and sometimes becomes brighter. Colors of the other cloud belts and spots may be caused by similar chemical action – by such substances as ammonia.

In addition to its many moons, Jupiter is circled by a ring system, similar in nature to the well-known rings of Saturn, but very much fainter, and invisible from Earth. Voyager photographed the ring by looking back at Jupiter with the Sun behind the planet. Illumination from behind shows up the ring at its brightest. The Voyager craft also photographed luminous streams of light known as auroras near the planet's poles. These are very much like the displays of light that are sometimes observed on Earth.

A Voyager 2 image of the Great Red Spot, taken in mid-1979. Within the spot a complex swirl of circulating cloud is the largest storm in the planetary system.

Jupiter's Moons

Jupiter controls a family of at least sixteen moons. The four largest of these are easily seen through a small telescope or binoculars. They are named Io, Europa, Ganymede and Callisto; and because Galileo was the first astronomer to see them when he turned his telescope toward Jupiter in 1610, they are often referred to as the Galilean moons. They look like small points of light. Sketches of the positions of Jupiter and its satellites made on several successive nights reveal the ever-changing pattern as the moons circle the planet, each in its own time. Io takes just under two days, Europa three and a half days, Ganymede a week, and Callisto almost seventeen days for one trip around Jupiter.

The Voyager missions discovered three new satellites of Jupiter, bringing the known total to sixteen.

I = Io   II = Europa   III = Ganymede   IV = Callisto

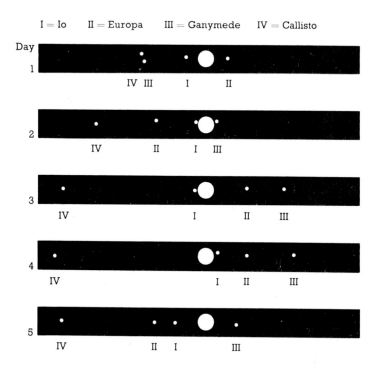

The four large moons of Jupiter can be followed from night to night with a small telescope; sketching their changing positions helps to identify them individually. The positions over a typical five-day period are illustrated here.

Further tiny moons almost certainly remain to be identified. The outer satellites measure, at most, a few dozen miles across. It is likely that they are asteroids captured by the planet's gravitational pull. The inner satellites may represent the original system that formed at the same time as Jupiter itself; the Galilean moons are among this group. Photographs taken by Voyagers 1 and 2 transformed these four anonymous pinpoints of light into real worlds.

Callisto has the most heavily cratered surface yet encountered in the solar system. The crowding of small craters suggests that Callisto's appearance has stayed comparatively unchanged for several billion years. It is a world with virtually no geological activity; and its density is so low that it must be composed of a mixture of rock and ice.

Scientists expected Ganymede, with its similar density and size, to look much the same as Callisto. But although parts of Ganymede are cratered, other regions of the surface are very different. For example, many straight, parallel lines of ridges and valleys suggest great activity inside Ganymede in former times, evidenced also by large craters, bigger than any on Callisto, with bright halos and rays.

Europa's smooth, icy surface is crisscrossed by numerous dark lines. Their nature is something of a mystery since they do not seem to be cracks, lower than the general level of the surface, or raised ridges. There are also some light-colored ridges, rising up to a height of several hundred feet, that trace out amazing scalloped patterns across the surface.

Io, the innermost Galilean moon, turned out to be the most startling. No less than eight huge and highly active volcanoes were spotted by Voyager 1 – some marked by plumes of incandescent gas spurting higher than 60 miles. Scientists estimate that 100,000 tons of sulfur and sulfurous gases are being spewed out every second. The riot of color on Io – red, orange and yellow – is explained by the hues of different forms of sulfur. The white areas are sulfur dioxide snow. Jupiter, the nearby giant, and the other Galilean moons, cause enormous tides within Io, which stir up the molten sulfur.

Satellites of Jupiter

Name and Number	Distance from Jupiter in miles	Orbital Period (days)	Year of Discovery	Radius (miles)
1979 J3	79,970	0.29	1980	12
Adrastea	80,500	0.30	1979	11
Amalthea	112,700	0.49	1892	81
1979 J2	200,000	0.67	1980	23
Io	262,000	1.77	1610	1128
Europa	417,000	3.55	1610	971
Ganymede	665,000	7.15	1610	1639
Callisto	1,168,000	16.70	1610	1497
Leda	6,925,000	240	1974	5
Himalia	7,130,800	251	1904	53
Lysithea	7,395,000	264	1938	11
Elara	7,294,000	260	1905	25
Ananke	12,200,000	630	1951	9
Carme	14,000,000	692	1938	12
Pasiphae	14,600,000	739	1908	16
Sinope	14,700,000	758	1914	11

Jupiter has sixteen known satellites whose principal characteristics are listed at left.

Numerous volcanic features can be seen in this image of Io's disk. The large darker areas are where debris has been widely scattered.

Callisto's extensive ring complex, known as Valhalla, is similar to the large circular impact basins that are visible on the Moon.

55

Saturn

Saturn is in many ways like a slightly smaller version of Jupiter, except that bands and markings are seen less often and are always indistinct. However, if the face of Saturn displays little to interest Earthbound astronomers, the magnificent system of rings is compensation. These can usually be seen easily with a small telescope. The rings lie at an angle to Saturn's orbit around the Sun. Because of this tilt, the rings, as viewed from Earth, sometimes appear open and sometimes edge-on. Twice during Saturn's twenty-nine-year journey around the Sun we see the rings open to the fullest possible extent, so that they appear very bright. Halfway between these wide open positions the rings can be seen edgeways; but except through powerful telescopes, they seem to vanish, because they are only a few miles thick.

Close-range photographs of Saturn taken by the Voyagers in 1981 indicate that the planet does have turbulent cloud belts and spots, similar to those of Jupiter, but the color contrast is not so marked, and they are veiled by haze high in Saturn's atmosphere. In fact, the wind speeds recorded for Saturn are up

A section of the numerous rings around Saturn is shown above. Within the major ring systems there may be hundreds, even thousands, of separate rings, composed of rock, ice and dust particles.

This image of Saturn, computer-processed from Voyager 2 data, reveals a banded structure of wind circulation in the upper atmosphere. It is immediately clear that the cloud systems on this planet are not nearly so prominent as on Jupiter.

Voyager 1 recorded Saturn and its glorious rings from a distance of about 12 million miles.

to four times faster than on Jupiter. Like Jupiter, Saturn is slowly shrinking and losing heat from its interior – an indication that the composition and structure of these two giants are also similar.

The multiple rings of Saturn are composed of numerous individual particles. Many of these fragments are probably little more than granules of ice and dust, but there are also much larger chunks of rock, perhaps up to 30 feet across. Viewed from Earth, the rings appear smooth, without any structure, apart from dark gaps. The Voyager pictures of the rings, however, revealed something quite different. The rings are made up of many hundreds of narrow ringlets, rather like the grooves of a record. There are mysterious spokelike features in the bright B-ring. These spokes seem to rotate with the planet as it turns on its axis. They may be created by Saturn's magnetic field.

The Voyagers took photographs of the rings with the Sun lighting them from behind. Such views cannot be seen from Earth. Under these lighting conditions the rings look quite different from the way they are usually seen. For example, the conspicuous B ring disappeared completely, revealing that particles in the ring are dense enough to block the sunlight. The major gap in the ring system, called the Cassini Division – dark when viewed from the Earth – shows up bright in the Voyager pictures, revealing that it actually contains material that can scatter sunlight.

The narrow F-ring consists of five strands, each 40 to 60 miles wide. The material in these strands is kept within the ring by the gravitational action of two small satellites discovered by Voyager 1, one of them orbiting just inside and the other just outside the F-ring. Nicknamed "shepherd" satellites, these bodies, along with large particles within the ring, may cause the strands to intertwine at certain segments, giving the ring a braided look.

Saturn's Moons

As a result of the Voyager encounters with Saturn, eight new moons were added to the planet's previous list of nine, and there may well be more. Five of the seventeen moons have diameters of more than 600 miles. Titan is by far the largest. It used to be considered the biggest satellite in the solar system, but is now known to be slightly smaller than Jupiter's Ganymede.

Most of Saturn's moons are known to contain a great deal of ice that reflects sunlight, making their surfaces appear bright. Phoebe – thought to be a captured asteroid – is probably composed entirely of stone. There may be some ice in its makeup, but its surface does not reflect as much sunlight as icy moons do.

Mimas, Enceladus, Tethys, Dione, Rhea, Hyperion, Iapetus and Phoebe have cratered surfaces as do the minor moons recently discovered.

Mimas is scarred by a crater 80 miles across – one-third of the satellite's diameter. Had the object that crashed into the surface to form this crater been just a

Satellites of Saturn				
Name and Number	Distance from Saturn in miles	Orbital Period (days)	Year of Discovery	Radius (miles)
1980 S28	85,000	0.6	1980	12
1980 S27	86,000	0.61	1980	43
1980 S26	88,000	0.62	1980	34
1980 S3	94,000	0.69	1966 and 1980	43
1980 S1	94,000	0.69	1966 and 1980	68
Mimas	115,000	0.94	1789	122
Enceladus	148,000	1.37	1789	155
Tethys	183,000	1.89	1684	325
1980 S13	183,000	1.89	1980	11
1980 S25	183,000	1.89	1980	11
Dione	234,000	2.74	1684	350
1980 S6	234,000	2.73	1980	111
Rhea	327,000	4.52	1672	475
Titan	760,000	15.95	1655	1600
Hyperion	920,000	21.28	1848	127
Iapetus	2,200,000	79.33	1671	450
Phoebe	8,000,000	550.00	1898	68

Enceladus is an intriguing object. Much of the surface is covered in ice and strongly reflects sunlight, making it the brightest of Saturn's moons. It is also the most youthful. The smooth plains at the right are only some 100 million years old.

The table at left shows the properties of the seventeen satellites of Saturn now known to exist. Information about the most recently discovered satellites was relayed to Earth by Voyagers 1 and 2.

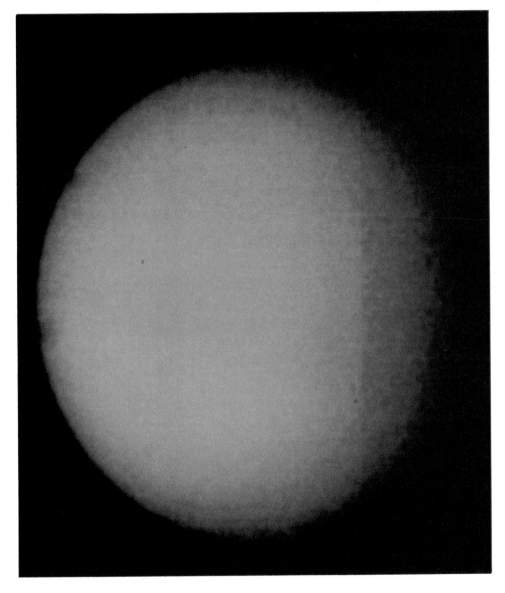

Mimas (above) has a heavily cratered surface. The conspicuous impact crater is about 80 miles in diameter.

Thick layers of gas and dust in Titan's atmosphere prevented Voyager 1 from scanning the surface of the second-largest satellite in the solar system.

little larger, Mimas might have disintegrated. Hyperion has a very irregular disklike shape and many craters. Perhaps pieces of Hyperion were knocked off by the impact of large meteoroids long ago.

From Earth-based observations, astronomers knew that Titan is the only moon in the solar system that has an atmosphere. So they awaited Voyager pictures of this satellite with special interest. However, Titan's atmosphere contains thick layers of reddish cloud, which makes the surface completely invisible. According to the Voyager instruments, the atmosphere proved to be chiefly nitrogen. Methane, and related gases, together with ammonia, make up the rest. Scientists consider that this atmosphere may provide insights into the properties of the Earth's own atmosphere billions of years in the past, before the dawn of plant life.

The Outer Solar System

Beyond Saturn are the two giant planets, Uranus and Neptune, as well as the smaller icy worlds of Pluto and its satellite Charon. These planets were unknown in ancient times and they have never been visited by spacecraft. The great German astronomer William Herschel, who worked in England, found Uranus in 1781 during a visual survey of the heavens. He wished to name the new planet after his patron, the reigning King George III, but eventually the planet was called Uranus. In classical Roman mythology Uranus was the father of Saturn, who in turn fathered Jupiter.

The planet Neptune was located in 1846 after some remarkable detective work by two mathematicians. Only ten years after the discovery of Uranus, astronomers noted that it failed to keep to the orbit predicted on the basis of Newton's law of gravitation. By 1830 they speculated that the deviations from the predicted path might be caused by the gravitational pull of an unknown planet. A young English mathematician from Cambridge, John Adams, and a Frenchman, Urbain Le Verrier, independently calculated the position of this unseen planet. Adams had no luck in persuading the Cambridge professor of astronomy to make a quick search. Instead, the Berlin Observatory worked on Le Verrier's calculations and finally tracked down the planet.

The finding of Neptune, named after the mythological god of the sea, did not resolve all the problems of the planetary orbits, so observers searched for a yet more remote planet. Percival Lowell, a Bostonian, led a systematic search for the planet until his death in 1916. Other scientists continued the search and finally discovered Pluto from the Lowell Observatory, Arizona, in 1930. Pluto, in Greek mythology, was lord of the underworld, but the name was given to the new planet because it includes Lowell's initials as its first two letters.

Planetary scientists now think that there cannot be another large unknown planet in our solar system. If this so-called Planet X really existed, we should by now be well aware of its gravitational force acting on the other planets. Of course, there may be countless objects a long way off. Other drifters may be out there, too tiny and too distant to show in our telescopes. In 1977, for example, a rock measuring about one hundred miles across was noticed drifting beyond the orbit of Saturn. This far-flung asteroid is named Chiron.

Uranus and Neptune are both about four times

Pluto (arrow) is so distant that on a photograph it looks like a star. But it is a planet, moving gradually against the distant starry background, as shown in this pair of photographs taken a few days apart.

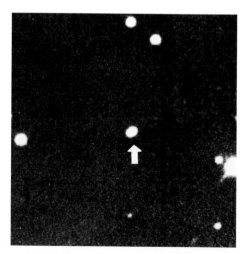

Uranus has five known satellites, seen in this photograph, all difficult to study because they are faint and close to Uranus. Their names, beginning with the one that orbits nearest the planet, are Miranda, Ariel, Umbriel, Titania and Oberon.

This computer simulation shows the encounter that will take place at 7 p.m. (Greenwich Mean Time) on January 24, 1986, when the Voyager spacecraft will be closest to Uranus.

larger than Earth, and each possesses an atmosphere of hydrogen, helium and methane. Uranus has five known natural satellites, and Neptune has two. A very odd feature of Uranus is that it lies on its side, with the rotational axis tipped over at an angle of 98°. This means that the seasons on the planet must be strange: for many years the Sun does not shine at all in one hemisphere, while the other is bathed in sunlight.

In March 1977 Uranus passed across the sightline to a distant star. Telescopes at observatories around the Indian Ocean and in a high-flying aircraft followed the planet's progress. To the surprise of the astronomers, the starlight was blotted out several times before the planet itself cut through the line of sight.

This unexpected event showed that Uranus has a system of narrow rings, now known to number about nine, each nearly filled with material.

Pluto is by far the smallest of the planets, an icy world with a diameter of about 1500 miles and a mass about one-tenth that of the Moon. Its orbit is an eccentric ellipse that at times places the planet inside the orbit of Neptune. Since 1979 Pluto has been closer to the Sun than Neptune is, a situation that will continue until 1999. It will reach its greatest distance from the Sun in 2113.

Pluto has one known satellite, Charon, discovered in 1978. It has a diameter of about 500 miles and orbits only about 10,000 miles above the planet.

Comets

Comets – chunks of frozen gases, water and dust – blaze their spectacular trails through space from the outermost parts of the solar system, far beyond the orbit of Pluto. The gravitational pull of mighty Jupiter traps some of these comets into closed orbits as they swing through the skies. They then return at regular intervals as periodic comets. The best-known example is Halley's Comet, named after the seventeenth-century English astronomer Edmund Halley, who computed the comet's orbit of 1682 and established it as a periodic comet – the first to be so identified – that would return in 1759.

Halley's Comet returns about every seventy-seven years and indeed it has been sighted on such occasions for the past two thousand years. This comet is depicted on the Bayeaux Tapestry, recording the Norman invasion of England of 1066, and in an Italian fresco of 1304. Astronomers next expect Halley's Comet to make its closest approach to the Sun early in February 1986. Three space agencies, those of Europe, the USSR and Japan, plan to launch spacecraft to fly by the comet, which should be visible from November 1985 to April 1986. The comet's track in Earth's sky means that conditions for observing will be best in the southern hemisphere.

Comets are normally named after their discoverers; and many of the new ones are found by amateur astronomers who specialize in searching for them. Several people might chance to see the same new comet within a few hours, but only the first three to report the discovery are entitled to have their names attached to it – for example Comet Tazo-Sato-Kosaka, first seen in 1969 and named for its Japanese discoverers.

Astronomers know of more than 600 individual comets. About half a dozen pass through the inner solar system each year; only once every year or so does a really brilliant comet move across the skies. Comets do not generate light; rather they reflect sunlight. When a comet first comes toward the Sun, it may look like a fuzzy patch of light. As it gets nearer to the Sun the comet becomes brighter. A comet may grow a shimmering tail, consisting of dust and gas that streams off as the Sun's energy boils its icy coating. As the comet travels away from the Sun, its head freezes up again and its tail disappears.

Much remains to be learned about comets. They may be remnants left from the early formation of the solar system; if so, they are a valuable sample of the material from which the Sun and the planets condensed. Astronomers have never seen the central part of a comet and can only guess as to its structure; they are eagerly awaiting 1986 to see what new facts the flyby spacecraft reveals about Halley's Comet.

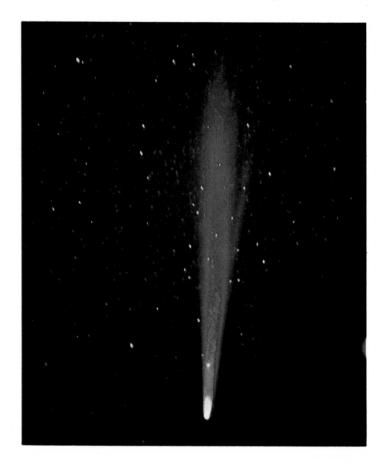

Comet Halley, photographed at the Lowell Observatory, Flagstaff, Arizona on March 19, 1910. The image has been computer-enhanced by the picture processing laboratory at the Kitt Peak National Observatory.

As a comet moves on its orbit around the Sun, the tail changes in size and direction. It always points away from the Sun, no matter in which direction the comet travels; the pressure of energy from sunlight and the gaseous solar wind push the tail away from the Sun. As the comet gets closer to the Sun, heat causes icy outer layers of the comet to evaporate, releasing gas and dust. In this way the comet grows a tail that can be as long as 100 million miles.

In this time exposure, the comet Humason, discovered in 1961 and last seen and photographed in 1962, appears at center right. With an elongated head several thousand miles in diameter, Humason – named for the American astronomer Milton Humason – came within 200 million miles of the Sun.

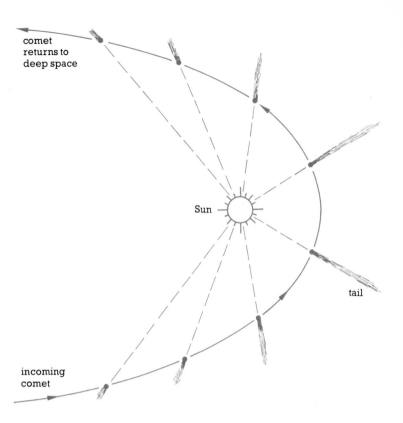

comet
returns to
deep space

Sun

tail

incoming
comet

The Sun

The Sun is an ordinary member of the starry skies. It is just one single star out of the 100 billion stars that make up the Milky Way. From a distance of many billions of miles, the Sun would look like any other common star.

To the plants, animals and peoples of the Earth, the Sun is a unique and vital star. Every living thing on the Earth owes its existence to the fact that the Sun is nearby and keeps shining, and has done so for about five billion years. The energy from the burning of coal, oil and natural gas was once sun-energy. These fuels are the remains of plants and animals that grew in the warmth of sun-energy millions of years ago. The nearest star – apart from the Sun – is 300,000 times farther away, and the weak star-energy we receive from it cannot possibly replace sun-energy.

The Sun is far larger than the Earth and also a great deal more massive. One hundred and nine Earth-planets placed side by side would stretch from one side of the Sun to the other. Its volume is 1.3 million times greater than the Earth and the mass 330,000 times as much.

The distance from Earth to Sun is about 93 million miles. Light and heat take eight minutes and twenty seconds to race across interplanetary space and reach the Earth from this distance. Although this seems a great separation, only a handful of stars exists within a million times this distance from the solar system.

The Sun's gravity pulls much harder than the Earth's gravity. A person who could venture to the surface of the Sun would weigh about one and a half tons. However, this is an impossible adventure since the Sun has no solid surface and the temperature there is about 10,000°F. This exceeds the melting temperature of every known substance. The temperature of the surface seems high, but inside the Sun it is much hotter. Its entire globe is a glowing mass of gas. At the center the temperature is about 27 million degrees Fahrenheit.

As little as a century ago scientists firmly believed that the Sun was nothing more than a flaming ball of fire. A sun made of blazing coal could not, in fact, last even for a million years before becoming a heap of ashen dust. Geologists have shown that the Earth is billions of years old, and that the Sun has shone throughout that time. In the 1930s physicists showed that the Sun and stars are powered by thermonuclear reactions.

This telescope at the Kitt Peak National Observatory in Arizona is the largest solar telescope in the world. The diagonal tower runs 280 feet underground to a 60-inch mirror that reflects the Sun's image to a smaller mirror at ground level. This mirror directs the image onto equipment that analyzes the Sun's composition, movements and magnetic field.

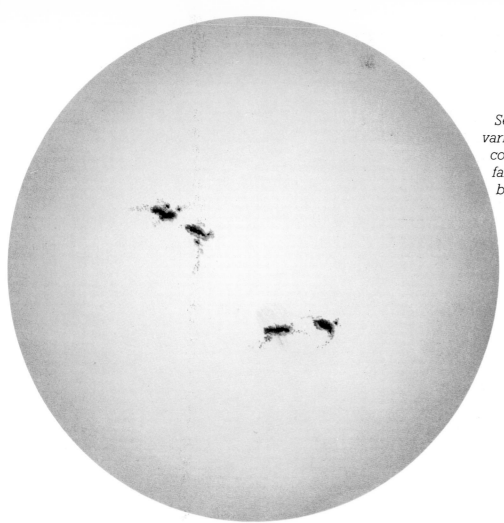

Seen through a telescope, the Sun has a variety of surface features. Dark sunspots come and go; flaming arches of gas leap far into space; and solar flares may flash brilliantly for up to an hour in the active regions. The visible yellow surface is called the photosphere. It marks the transition between the immensely hot and opaque interior and the thin transparent atmosphere that allows the Sun-energy to stream into the blackness of space.

The gas inside the Sun is three-quarters hydrogen, the lightest gas. Deep inside the hot Sun, hydrogen atoms crowd together. In the jostling a group of them collides so violently with another group that they fuse together and make a completely different substance, helium. Each second, 650 million tons of hydrogen become helium. A small part of this mass of material is transformed in the process and reappears as pure energy, as Einstein had predicted would be the case. In one second, the Sun's mass falls by four million tons. In fifty million years the lost mass is equal to the mass of the Earth.

Flashes of energy burst forth as the hydrogen turns to helium. The great density of matter traps the energy flashes inside the Sun. They wander through the interior for a million years or so before reaching the surface. The energy then streams off into space.

Along with the heat and light, the Sun emits radiation that can be harmful to living creatures. Ultraviolet rays and X-rays damage the cells in plants and animals. The Earth's blanket of atmosphere soaks up almost all of this radiation, although the small amount that reaches the ground on a fine day will make fair skin tan, or cause painful sunburn if exposure is too long. Astronauts journeying into space have to be protected from the Sun's harmful rays.

The sun's blinding light will hurt your eyes permanently if you stare at it. **No one should ever look at the Sun through any type of magnifier, field glasses or a telescope.** Nor should you use the cheap filters sometimes sold with small telescopes.

Even astronomers do not look through their telescopes directly at the Sun. Special instruments fitted to solar telescopes allow them to follow the behavior of the Sun by indirect viewing systems that protect their eyes.

Spots on the Sun

The yellow disk of the Sun is actually the topmost layer of the star's glowing gas. Above this layer is the cool zone of transparent gas called the chromosphere. Solar astronomers use photographs taken with special telescopes to study the gaseous surface. Frequently, the photographs are made with filters that isolate the light associated with one particular type of atom, such as hydrogen. Photographs made in this way allow astronomers to study different layers of the Sun's outer surface.

A photograph taken under very good observing conditions shows that the Sun's surface has a mottled or bubbly appearance. Glowing bubbles of gas jostle, with darker spaces snaking between them. New flecks continuously appear, and then, after a few moments, dissolve away again. Each of these bubbles is generally about as large as Florida. They are, in fact, gushers of hot gas shooting up from just inside the Sun. The yellow layer is somewhat like the surface of a boiling pan of syrup.

Dark blotches called sunspots were noticed even in ancient times. The largest groups of sunspots may stretch across 190,000 miles which is nineteen per cent of the Sun's visible width. These giant sunspots are rare, but they can be seen by the naked eye at sunset if they are present.

Do not try to look directly at the Sun or its sunspots with any magnifier or telescope. Eye damage may result. Instead, with a simple telescope on a sturdy tripod, it is possible to project an image of the Sun onto a white card. A little practice will produce a sharp image of the Sun. Sunspots will appear as grayish specks. A record of the positions of the spots on several successive days will show that sunspots change in size and shape, and that the Sun itself slowly rotates.

Sunspots look like holes in the fiery surface of the Sun. In fact they are areas that are about 3000°F cooler than the surrounding surface. This makes their temperature roughly 7000°F. Something that hot is actually extremely brilliant: sunspots only *look* dark because they are cooler and dimmer than the rest of the Sun. If a sunspot could be plucked from the Sun and examined separately it would seem a hundred times brighter than the full Moon.

How are sunspots caused? The Sun's very strong magnetism probably bursts out from the inside of the Sun, and a pair of spots is created at those two points where the magnetism leaves and subsequently re-enters the Sun. Small spots vanish in a matter of hours but the larger ones may remain for several weeks before finally disintegrating. An average spot is 20,000 miles across; most spots are more or less as big as the Earth, and huge spots span 90,000 miles.

Records of sunspots for about 300 years show that the number of spots and their size varies on a cycle lasting for about eleven years. The number of spots steadily increases for five or six years; then there is a decline in numbers for the following four or five years.

An astronomer projects the image of the Sun onto a card held beyond the eyepiece of a telescope. Even on the lowest magnification, small spots can be seen by this safe method of viewing the Sun. A cardboard shield resting on the eyepiece casts a shadow on the lower card; otherwise the direct sunlight would blot out the image on the card. The finder telescope has been removed from the main tube to eliminate the temptation of looking through it – and damaging eyesight – while lining up the instrument.

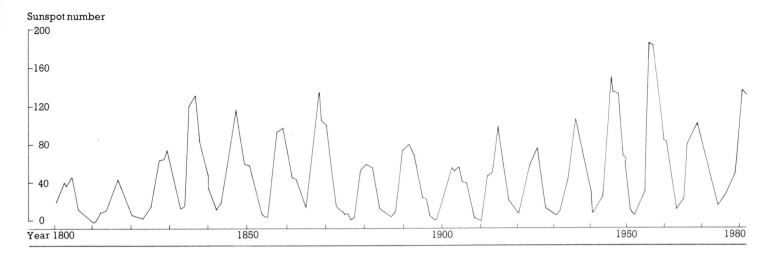

Sunspot number

Year 1800   1850   1900   1950   1980

The number and size of sunspots fluctuate on an almost regular cycle. The sunspot number (a measure of sunspot activity) is here plotted as a yearly average. Highs and lows are spaced at roughly eleven-year intervals. It is possible that this sunspot cycle may influence the weather on the Earth.

At the end of the cycle, there may be no spots at all for many months until the new cycle gets going. Vast changes in the Sun's magnetism probably cause this variation in the sunspot numbers. The last cycle peaked to maximum in 1981 and the next is predicted for 1992–3. Many other aspects of the Sun's behavior change through the cycle, and this probably affects the weather on the Earth.

Sunspots can interfere with radio communications. Electric particles are flung into space by solar storms known as flares, and eruptions near the sunspots. The solar electricity stirred up in these storms changes the upper part of the Earth's atmosphere. At such times, fadeout of long-range radio signals may be noticed. In 1980 extremely sensitive measurements showed that the Sun's energy output varies slightly from day to day. Possibly the changing pattern of spots and flares causes this small variation.

The chromosphere, the cool layer of atmosphere above the yellow surface, can be seen readily only during total eclipses. The temperature in this thin layer is about 8000°F. Above the chromosphere is the intensely hot and invisible corona, where the temperature soars to an amazing two million degrees Fahrenheit. The gas in the corona is boiling away into space. This gas rush is called solar wind.

A solar storm rages close to sunspots, producing flares of white light, as particles of electricity shoot off into space.

67

Action on the Sun

Great storms occasionally erupt in the Sun's atmosphere. Extra-hot gases swirl and boil up from the interior and then stream high above the surface. Near sunspots, flares may burst out. A flare is a spectacular release of energy in the chromosphere. A vast region switches on and off in a series of dramatic flashes, each one like an enormous lightning stroke. In about thirty minutes the flare spends all its energy and then subsides.

The flaring activity flings large clouds of electrically charged particles away from the Sun. These travel at up to two million miles per hour. After about two days, the particles reach the vicinity of the Earth. Some of them are funneled in toward the polar regions of the Earth. This is because the Earth is magnetized in the same manner as a giant bar magnet and the electric particles from the Sun are guided by the magnetism. The magnetism is strongest near the North and South Poles. As the charged particles are drawn down into these regions, they strike the gases in the upper part of the atmosphere. This makes the air glow and emit the beautiful light displays known as auroras.

The finest auroral displays take place two years after maximum sunspot activity. In periods of particularly intense bombardment by solar particles, auroras can be seen over southern Europe or the southern United States. Normally, however, it is only seen farther north. A flight in an airplane across the Atlantic Ocean at night may provide a chance to see stunning displays as the aircraft passes close to Greenland. In Australia the southern aurora is rarely seen.

Earth is equipped with an invisible shield of magnetic force, protecting its surface from the worst blasts of high-speed particles ejected from the angry face of the Sun. This magnetism forms a barrier around the Earth that deflects most of the particles or funnels them in at the Poles. Inside the magnetic barrier, called the magnetosphere, two doughnut-shaped compartments can trap electric particles. These rings are

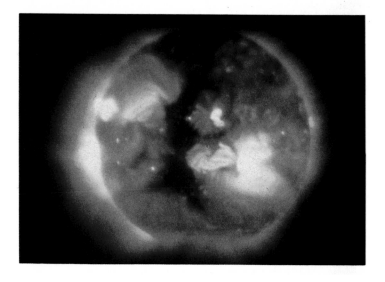

A Skylab X-ray photograph of the Sun's corona reveals a gaping hole in the coronal gases. From such regions the solar wind streams into space.

This computer-processed, ultraviolet light image shows disturbances in a section of the corona above an active region of the Sun.

An aurora, such as the one at left illuminating the treetops of a northern forest, produces richly colored curtains of light that seemingly dance in the sky. The aurora and its colors are triggered by collisions between electrically charged particles ejected from the Sun and the upper layers of the Earth's atmosphere.

named the Van Allen Belts after their discoverer, the American physicist James Van Allen.

During a total eclipse another type of solar disturbance is seen. Fantastic plumes of gas arch their way above the Sun, reaching altitudes of hundreds of thousands of miles before crashing back to the surface. Called prominences, they can be seen by special solar telescopes at any time.

X-ray telescopes on satellites can photograph the corona of the Sun, which at its temperature of two million degrees Fahrenheit emits X-rays rather than visible light. The corona has tangles of magnetism running through it. These zones may show as bright spots on X-ray photographs.

In X-ray images the visible surface of the Sun appears black, and only the coronal regions can be seen. It has been found that the corona is made almost entirely of loops of gas, arching their way around the Sun. Within the corona there are also enormous holes, black regions on the X-ray photographs, which are a source of the streams of gas and particles sent out by the Sun in the solar wind.

Shown in this schematic painting, two doughnut-shaped radiation belts encircle the Earth. Farther out from the Earth is the magnetosphere, which shields the planet from the solar wind.

The Stars

Each twinkling point of light in the night sky is really a distant sun. Even through a telescope the stars still appear merely as points of light, although with a telescope many more stars can be detected than can be seen by eye alone. But the stars are so much farther away from us than the Sun that even the world's largest telescopes are unable to discern the surface features of the nearest ordinary stars. Only in the case of a handful of giant stars can the telescopes vaguely make out the disk-shape of the star. Sometimes star images in telescopes, or on photographs, appear like disks, but in this case the disk is a distortion of the ideal image – a point of light – caused by the telescope and the Earth's atmosphere.

The nearest star to the solar system, Proxima Centauri, is more than four light years away. A light year is the distance covered by a light ray in one year. Since light travels at 186,000 miles every second, there are six trillion miles in a light year. It is a useful unit for marking out large distances in astronomy.

The distances to the nearer stars are found by watching the way in which they change their positions slightly against the background of more distant stars during the course of the year. This relative measurable change of position is called parallax. A passenger on a speeding bus or train experiences the parallax effect when looking out of the window. Nearby objects flash past the window, whereas objects farther away can be seen for some time. The parallax effect causes the nearby stars to shift their apparent position across the sky as the Earth makes its annual journey around the Sun. The parallax method only works for finding distances to one hundred light years. Beyond that distance, astronomers have devised ingenious methods for finding distance. Many of these involve a careful analysis of the light from the star to determine its properties. Certain types of variable stars, for example, fluctuate in a manner that directly indicates intrinsic brightness. Such stars are, in a sense, showing how bright they really are. By comparing this intrinsic brightness with the apparent brightness, an astronomer can calculate a variable star's distance.

The stars are not scattered evenly throughout all of space. The Sun and all the stars that can be seen by eye belong to a giant disk-shaped collection of a hundred billion stars in the Milky Way. The Galaxy is 100,000 light years from side to side. Beyond its edge space is totally devoid of stars until neighboring galaxies are reached.

Some of the stars in the Galaxy are alone, like the Sun, but many are linked in double, triple, and quadruple systems. Stars also form larger groupings called clusters. Two quite distinct kinds of cluster are found: open clusters and globular clusters. The open clusters are rather loose collections of stars, about one thousand in all. Many individual stars can be picked out in an open cluster. For example, The Pleiades (Seven Sisters) in the constellation Taurus is a good group to look at. Several of the stars can be

Astronomers have estimated that this swarm of stars, making up the globular cluster in the constellation of Centaurus, has looked this way for billions of years.

seen without a telescope, although the misty patch of light becomes a hundred twinkling stars when seen with field glasses. Globular clusters are different from open clusters. They are tightly packed balls of stars, with up to a hundred thousand members. In this swarm, only stars at the edge can be picked out individually.

Stars do not appear to be equally bright in the night sky. The range of star brightnesses is due partly to the fact that the stars are scattered and not all at the same distance away, and partly to the fact that they do not all give out the same amount of light. There are stars of many sizes, temperatures and masses, and they vary, too, in their colors. The most massive stars contain enough material to make about fifty Suns, while the smallest have only a few tenths of the Sun's mass. The most massive stars are also the hottest and

they are large, shining blue-white. The yellow Sun is a smallish star, but some cool, red stars are even smaller. Small stars are often called dwarfs. The type of star known as a white dwarf is less than one-hundredth the size of the Sun. In contrast, there are also giant stars that have puffed out to become many times larger than they once were. Really immense stars are called supergiants, and they are several hundred times bigger than the Sun.

The overall patterns of the stars in the sky will not change noticeably in a human lifetime, yet the stars are in motion. We can barely detect the star movements even with telescopes because they are so far away. Very careful measurements have to be made on photographs taken a few years apart. Such measurements show that the Milky Way is a great swirling mass, a disk of stars turning slowly in space.

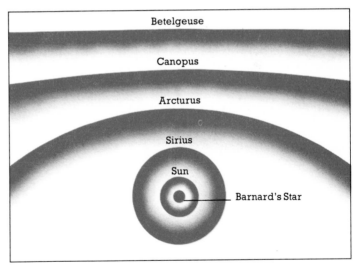

The diagram above gives an idea of comparative star sizes. Barnard's star is a faint dwarf star six light years from the Sun. Betelgeuse is 500 times the size of the Sun.

The open star clusters in the constellation of Perseus formed about five million or so years ago. Each of these families has a few dozen members. The Sun was once a member of such a cluster.

The Message of Starlight

Though a single speck of starlight may seem insignificant against the blackness of the night, each tiny beam is packed with information about its sender. Astronomers have learned how to read the messages in starlight, which reveals what the stars are made of, how big and hot they are, and how they are moving.

Before the information in starlight can be understood, it has to be decoded. First of all the light from a star is collected and concentrated by a telescope. Next, a special instrument on the telescope analyzes the light. The instrument is called a spectrograph. The result is stored as a photograph or on computer tape or magnetic disk.

Light is often described as consisting of "light waves." Light does indeed travel in the form of waves, but the distances between the wave crests are so small that thousands fit into a millimeter. Light waves are energy waves and they can travel through the total emptiness of space. The eye can see things that give out their own light energy, such as the Sun, the stars or an electric light bulb. The eye can also see objects that reflect light, such as the planets. Light waves come in a whole range of sizes. Some are longer, some shorter. The difference between light waves of different sizes is discernible because the eye sees them as different colors.

Different mixtures of light waves offer an infinity of shades from which to choose. The hues of the rainbow show how waves of different sizes have different colors. Red light has the longest waves, and

A beam of white light, containing a mixture of light waves of different wavelengths passes through a glass prism. The prism has the effect of splitting the beam into the various colors making up the light. The result is a familiar band of rainbow colors.

violet light the shortest. A rainbow is caused when the Sun's light shines through falling raindrops. The blobs of water are able to split the sunlight up into its range of colors – into what is called a spectrum (the plural of spectrum is spectra).

A spectrum can be made from any beam of light. A prism – a piece of glass or quartz used in optical experiments – in the path of the light beam will split it up into its rainbow colors. Another way of getting a spectrum is to use a piece of flat glass with many thousands of fine lines engraved close together. This is called a diffraction grating, and it is an essential tool of modern astronomy.

Inside an astronomer's spectrograph, a prism or a grating splits the starlight into its range of colors. The colors are in varying proportions from the different kinds of stars. Cool stars, for example, are sending out more red light than blue or violet. That is why they look reddish in the sky. Hot stars emit light more evenly through the spectrum. All the colors combined look white or bluish.

With some particular colors there are narrow gaps in a star's spectrum where there is very little light.

The gaps look like black lines cutting across the continuous background rainbow. They are called absorption lines. These lines are very important because the numbers and positions of the lines reveal what elements are present in the stars, such as hydrogen. Different gases absorb different colors as the light leaves a star. Each type of element puts its own "fingerprint" of black lines into the spectrum, and this allows particular elements to be identified. From the study of absorption lines astronomers have learned that stars are made mainly of hydrogen. The rest of the material, about a quarter, is helium. Many other elements such as oxygen, silicon, iron, and nickel make up only one per cent of a normal star.

Human beings have evolved on the Earth, bathed in sunlight. As a result their eyes are most sensitive to the range of light colors emitted by the Sun. However, there are other kinds of "light," which human eyes cannot detect. Though each kind of radiation has been given a different name, they are all really like light, except that each has waves of a different size. Gamma-rays, X-rays, ultraviolet, infrared, microwave and radio waves are all part of the same family of radiation. All these types of rays can be picked up from various objects in space, but special telescopes have to be used.

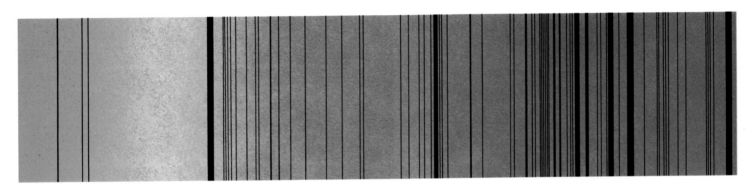

The color spectra of the Sun and stars, such as those shown above, are crossed by many narrow, dark absorption lines, often called Fraunhofer lines, after Joseph von Fraunhofer, who first listed some of them. They help to discover the atomic elements present in the Sun and the stars.

The telescope that took the photograph below of a cluster of stars had a prism placed across the front. Instead of appearing as a point of light, each star is drawn out into its spectrum. A close look shows fine dark absorption lines crossing each tiny spectrum.

Double Stars and Variable Stars

The Sun is a solitary star, over four light years from its nearest neighbor, but many of the stars in the sky are actually double. The two stars that make up a double, or binary, star are held together by the pull of gravity between them. In a binary star, each member is in orbit around a balancing point between the two, called the center of mass. Closely spaced pairs of stars may take only a day or two to complete their circuits. Distant pairs may take over a hundred years.

Some double stars can be detected easily with a telescope. Both stars can be seen, and over a period of time one seems to travel around the other. But only the pairs separated by large distances are directly observable. Close pairs cannot be split apart into two star images, even in the largest telescopes. However, astronomers can tell when there are actually two stars very close together by looking at a spectrum. As the stars circuit their orbits, there is a regular cycle of changes in the spectrum; rotation shifts spectral lines within the colors toward red and blue and back. In addition to helping identify the stars as a pair, the study of this spectrum makes it possible to find out

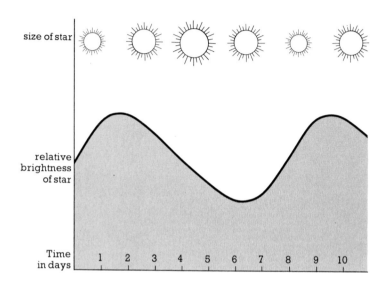

This variable star grows and shrinks as nuclear activity in it varies with a regular cycle. As its size changes, so does the brightness. Maximum brightness occurs before maximum size because the drop in temperature as the star expands has a greater effect on brightness than does the increase in size.

The members of an eclipsing binary system pass behind each other as they move in their orbits. The amount of light that is visible dips during each eclipse. In this example the two stars have different sizes and brightnesses. The bigger dip in brightness comes when the larger, fainter star blots out the smaller and more luminous star.

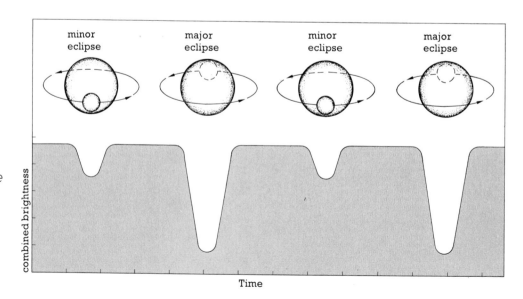

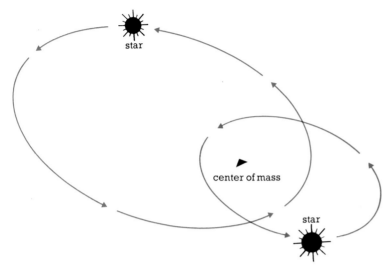

star

center of mass

star

In a double star system, each member travels in its own orbit around the center of gravity between them, called the center of mass. Each orbit is an ellipse. Like a pair of ice skaters holding hands and circling, the two stars always keep on opposite sides of the center of mass as they swing around.

what each individual star is like. Furthermore, by a careful study of the orbital motions within a binary star system, astronomers can sometimes calculate the masses of the individual stars.

Besides doubles, there are star systems with three or even more members, though not many are known. A famous example is Castor in the constellation Gemini. There are six stars altogether in this multiple system. Three stars can be detected with a telescope, but each, in turn, is a close double.

The orbits of some double stars are lined up so that from Earth each star disappears behind the other one as they move around in their orbits. Normally light from both stars is visible, but when one of them is hidden, there is a sudden dip in the amount of light. The starlight gets fainter for a short while, then returns to its original brightness. Pairs of stars like this are called eclipsing binaries.

There is a well-known eclipsing binary in the constellation Perseus. It is called Algol. Every sixty-nine hours the brightness of the Algol double star drops by more than a magnitude for a few hours.

Eclipsing stars are not the only ones whose light output varies. There are many other kinds of variable stars. Some of them change in a very regular way. The Cepheid variables, named after a star in the constellation of Cepheus, are an example. These stars actually swell and shrink quite regularly. As they expand and contract, their brightness rises and falls. Other variables are less well-behaved. Some that are usually dim unexpectedly flare up every now and again, gradually fading to their former level after each outburst. Others do the opposite and without any warning become fainter than they usually are. The variables are older, giant stars. They are well into the later part of their lifetime. The forces that normally keep stars like the Sun shining steadily have got out of balance, causing the stars to fluctuate in brightness.

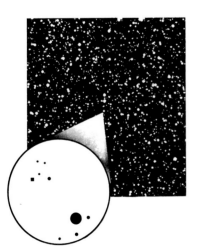 

A nova is a star that suddenly flares up to be many magnitudes brighter than before. At left are two photographs of the same part of the sky before and after a nova's outburst. The inset far left is an enlargement of the area indicated by the arrow, and the small square box at left in the circle shows where the nova appeared. The nova is very bright in the right hand picture. After flaring, a nova gradually fades again over a period ranging from a few months to years.

The Life of a Star

Over the human lifespan, the countless stars of the Milky Way seem practically changeless. Occasionally, a very faint star suddenly flares up, in what is called a nova outburst, to change the pattern of a familiar constellation for a few weeks, until once again it fades into insignificance. Rarely, even a greater outburst, called a supernova, blazes forth, like the one in 1054 AD that left behind the well-known Crab Nebula. Variable stars flicker with uncertain light.

The stars do eventually alter. Nothing in the Universe lasts forever. A star dies when its huge store of nuclear fuel is finally exhausted. Even today old stars are fading out, while new ones are being born to replace them. We can see stars in all stages of evolution, from infancy to old age.

Very young stars are found still embedded in the gas from which they form. The first glimmering light from brand new stars has been seen in the Orion Nebula, for example. The Sun is comfortably settled in middle age. Some of the oldest stars that are known are in the globular clusters.

How is it possible to work out the ages of the stars? Nobody can follow the progress of a single star from birth to death. However, imagine a person who has never before seen a tree and is suddenly transported to the middle of a forest. There he will see trees in every state of development from seedlings to gnarled giants. Knowing a little biology, he would soon work out the life cycle of a tree. In a similar way, from the laws of physics and observations of stars of different kinds, astronomers can deduce the sequence of events in the life of a star.

This illustration charts the changes that may occur during the lifetime of a star such as the Sun, from birth in a cloud of gas through advancing age. The star here expands to form a red giant, then decreases in size, eventually becoming a white dwarf.

After a star has formed, it soon settles down to a steady existence. Nuclear reactions in its innermost core convert hydrogen to helium, releasing energy at the same time. Eventually, all the hydrogen in the interior is consumed. Changes then take place in the star's internal balance. The outer layers puff out to give the star giant proportions, making it a red giant, while new reactions, working on the helium, start up inside. More changes occur and the star will go through a phase of being variable. Ultimately, there is no possible source of energy left. Smaller stars shrink into white dwarfs. Massive stars blow up as supernovas. The material blasted out by a supernova becomes part of the interstellar gas, the birthplace of a new generation of stars.

One of the last stages in the lives of stars, before they become white dwarfs, produces some of the most fascinating objects in the sky. These are the planetary nebulas (the plural form of *nebula*, a Latin word meaning a cloud). Their regular shapes and beautiful colors make them very attractive. In reality they have nothing to do with planets. The name is a relic from early telescopic observers who thought their disk shapes looked similar to planets. A planetary nebula is formed when the star at the center throws off a layer of itself, the shell of gas traveling outward.

The photographs at right show two of the 1000 or so planetary nebulas discovered by astronomers. The nebula at right has a characteristic shape that has earned it the name Dumbbell Nebula. In the exact center of the beautiful Ring Nebula, at far right, the dead star that threw off the ring can be seen clearly.

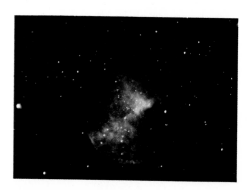

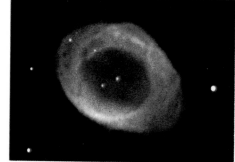

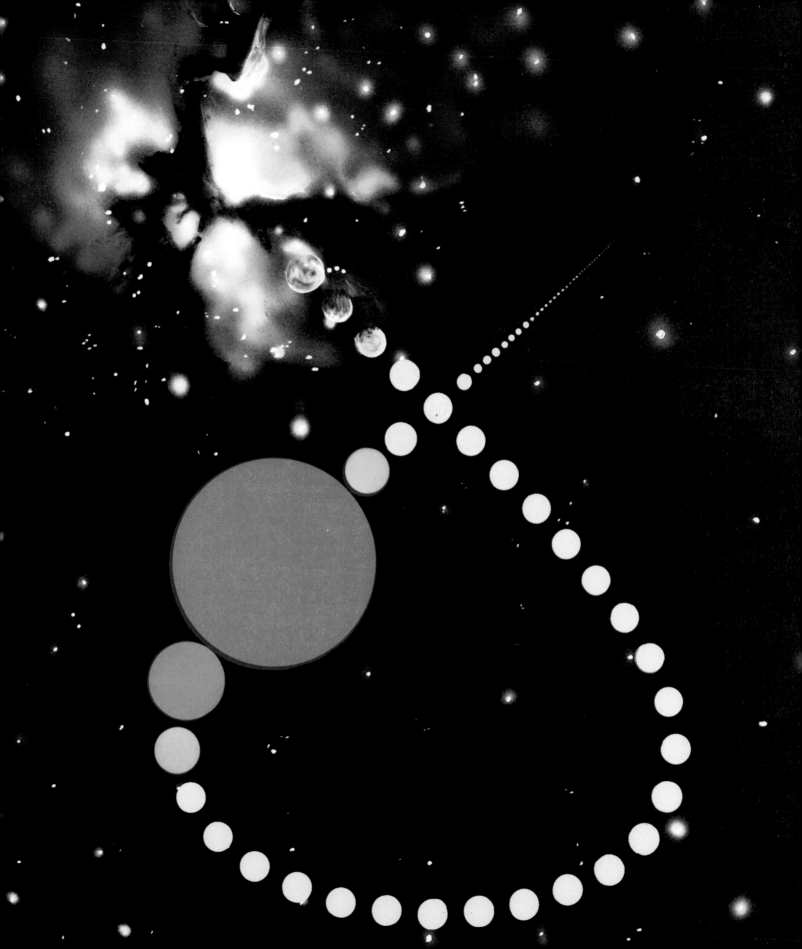

White Dwarfs and Black Holes

Throughout its life a star is a battle arena. The force of gravity tries to crush the star in on itself, but this inward shrinking is resisted by the outward pressure of the star's nuclear material. Eventually, however, the star becomes exhausted. Gravity takes control and the star takes on a form quite unlike that of normal healthy stars such as the Sun. A massive star may even disappear altogether by turning into a black hole.

The force of gravity always attracts: it wants to pull particles of matter closer together all the time. humans experience weight because the mass of the Earth tugs at the mass of a body. Every atom of a body is attracting all other atoms by gravity. Since a normal star is very massive, up to a million times more

This diagram compares the size of a normal star with those of a white dwarf, neutron star and black hole. The relative size of the black hole is hypothetical because astronomers have not yet been able to measure black holes accurately.

massive than Earth, its internal gravitation is high. Try to imagine what it is like inside the Sun: at one-tenth of the distance from the outside to the core the pressure is already a million times higher than atmospheric pressure at the surface of the Earth. Halfway in it rises to a billion atmospheres. This crush is resisted by the pressure exerted by the hot gas inside the Sun. The gas is kept heated by the nuclear furnace.

When the nuclear fires finally dwindle, the star gas cools and then gravity is in command. What happens at this stage depends on the mass of the star.

A dying star similar to the Sun collapses until it is about as large as the Earth. No really spectacular explosion occurs. It just falls into a heap of radioactive ash and gently flickers out. It has turned into a white dwarf star. On Earth a cupful of white dwarf matter would weigh one hundred tons.

If a star is somewhat more massive than the Sun, the inward plunge pushes it past the white dwarf stage.

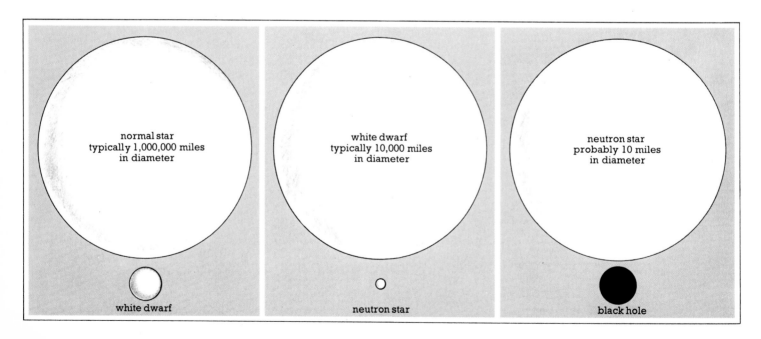

normal star
typically 1,000,000 miles
in diameter

white dwarf
typically 10,000 miles
in diameter

neutron star
probably 10 miles
in diameter

white dwarf

neutron star

black hole

The collapse does not stop until the star is ten miles across. At this point it has become a neutron star, a dense ball of nuclear particles. A cupful of neutron star material weighs nearly a trillion tons. Some neutron stars spin rapidly and emit a flash, or pulse, of radio waves on each rotation. This type of neutron star is called a pulsar. A neutron star will not form without first causing a stupendous explosion. In its last few minutes a dying star may explode as a supernova. For a few days it outshines entire galaxies. Then the central part of the supernova forms a neutron star.

It is not possible for a neutron star to be greater in mass than two Suns. A dying star of ten Sun masses, for example, suffers so badly from the gravity it generates that no force is large enough to withstand the collapse. When a star like this has shrunk to a couple of miles, its gravity is so huge that, as astronomers put it, the escape velocity is greater than the velocity of light. Anything – rockets, particles, light, or radio signals – trying to escape from the dying star cannot do so. Gravity is so strong that it drags everything back to the star. As far as the human observer is concerned, the star has become a black hole. It cannot be seen because it is so dense, with such powerful internal gravity that no light can escape from it.

The presence of black holes in space is revealed by the drastic effect they have on nearby matter. X-ray telescopes have detected a dozen or so exotic binary stars, which are very powerful sources of X-rays. In these strange binaries a black hole is probably orbiting close to a giant star. It tears at the outer layers of the giant, dragging them into a swirling disk of hot gas that hurtles down into the black hole. The disk around the hole glows intensely hot, which causes it to emit the strong X-rays.

Within the tangle of the Crab Nebula, a tiny neutron star (arrow) spins thirty times a second. The nebula and its neutron star are the remains of a star that exploded 900 years ago.

A supernova erupts in a distant spiral galaxy. For a few days a dying massive star shines out with the light of millions of Suns before dwindling to a feeble flicker. In the final long-exposure photograph the galaxy appears clearly, but the star that exploded has faded from view.

1937

1938

1942

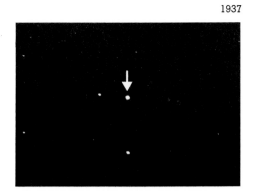

79

Between the Stars

Although the stars in the night sky look close together, they are separated by immense stretches of space. Most of the gaps between the stars look dark, but space is not in fact completely empty. There are atoms of gas and particles of dust floating throughout space. This matter creates a thin haze that dims the light from distant stars. Astronomers call the space dust and gas interstellar matter. It is much thinner than air. A cup of air contains about six sextillion atoms, but a cup of interstellar matter would contain only five hundred atoms.

In places the gas and dust have collected together, or been swept up by gravity, to form thicker clouds. Many of these clouds are so thick that they completely shut off the light from stars behind them.

Besides the dark dust clouds, there are also glowing clouds of gas, called nebulas, that emit pinkish light. These are among the most beautiful objects in the sky, and include the Carina Nebula and the Orion Nebula.

In winter Orion the Hunter is one of the easiest star groups to recognize. Just below the row of three stars that make up his belt there is a fuzzy patch of light. This is the Great Nebula in Orion, one of the few that can be spotted without a telescope. Field glasses or a small telescope will show that it glows with a soft light.

When gases are made to glow, each different gas gives out light of a different color. Everyone has seen the colored neon lights used in advertising signs; and sodium lights are frequently used to illuminate main highways. Neon is a gas that glows orange when an electric current is passed through it. Sodium lights look yellow.

In color photographs the shining nebulas of space usually look pinkish or purple. Most of the gas in space is hydrogen and the bright stars inside the nebulas cause the hydrogen to glow in a similar way to neon. Hot stars give out invisible ultraviolet rays that make the hydrogen gas shine with a beautiful pink-red light when the rays travel through it. As well as these bright clouds, there are other nebulas that act as cosmic mirrors, reflecting the visible light from stars that are near them.

An exciting discovery about the gas clouds of space is that new stars are being made inside them all the time. Astronomers have actually seen new stars turn on inside the Orion Nebula. A new star begins when particles of gas and dust collect into a huge ball. The pull of gravity squeezes the ball tighter and tighter, and at the same time it gets hotter and hotter. Eventually, the ball is hot enough to start the nuclear reactions, and it continues to shine of its own accord. Then it settles down to a lifetime of being an ordinary star, spending much of its time in an existence similar to the Sun's.

In the Milky Way – the great family of stars to which the Sun belongs – about nine-tenths of the material has already formed into stars. The other one-tenth is the gas and dust spread out between the stars. Within this material new stars are being created. The new stars are often grouped into open clusters, such as The Pleiades.

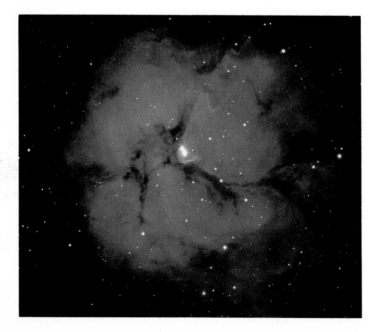

The Trifid Nebula is a dense cloud of hydrogen, laced with dark lanes of dust. Very hot stars deep inside the gas make it glow red. The nebula, which is 3000 light years from the Earth, can be observed through a small telescope.

In skies of the southern hemisphere the most glorious nebula is in the constellation of Carina. Many bright stars can be seen shining through this magnificent hydrogen cloud. At the center of the Carina Nebula there is a strange star, Eta Carinae, that changed greatly in brightness over twenty-five years during the last century. It was once the second brightest star in the southern sky but can now only be seen with a telescope.

A nebula called 30 Doradus lies in the Large Magellanic Cloud, a star family just beyond the Milky Way. This nebula is far larger than any in the Milky Way.

Dust Clouds and Life

Enormous clouds of space dust and cold gas, appearing as dark patches, cut through the silvery haze of the Milky Way. Near the Sun, the space dust is not noticeable. But stars far away, more than a thousand light years of distance, shine with a reddish hue. The Sun looks very red when it is close to the horizon; that is because its light is passing through murky layers in the atmosphere. Similarly, when starlight has to travel hundreds of light years through interstellar dust, it gets redder in directions where the dust is particularly thick. Stars tens of thousands of light years away can scarcely be seen, even through large telescopes. An optical telescope cannot see deep into the heart of the Milky Way for this reason.

The Horsehead Nebula in the constellation of Orion is a dense cloud of dark gas and dust. Behind the nebula brilliant young stars emit a flood of light that makes the horse's head stand out dramatically.

Sometimes the dust collects together with gas and forms a dark nebula, that is a cloud in space that emits no light. The most striking examples of this are The Coalsack, an astonishingly black void in the southern Milky Way, and the superbly shaped Horsehead Nebula in Orion. Infrared telescopes have shown that the youngest and earliest stages of star formation take place in the dark dust nebulas.

Within the Orion Nebula there is a large dust cloud that emits radio waves. Similar dust clouds occur in the constellation Sagittarius, toward the central section of the Milky Way. Radio astronomers have discovered that some of the signals from these dust clouds are caused by various kinds of chemical molecules.

Over forty different types of molecule are present in thick dust clouds. Some are simple substances, for example, carbon monoxide, which is made of carbon and oxygen. Others are more complicated, being made of several atoms; formic acid is an example – on Earth it puts the sting in nettles and bees.

All life on the Earth is based on the structure of one huge molecule named deoxyribonucleic acid – DNA. This molecule carries in code form instructions on how to duplicate itself. The molecular assembly line uses simpler components, such as sugar and protein, to build up DNA. These substances can be manufactured from even simpler molecules. In the dust clouds of space the simplest molecules needed to start the assembly of the complex substances of life exist in great profusion. Perhaps the first living things on Earth grew from these elementary molecules. It is within the densest clouds that new stars and planets are born. When the Earth first condensed it may have absorbed life-forming molecules from a cosmic dust cloud.

The beautiful Orion Nebula (right) is a birthplace of stars. Hot hydrogen gas flushes a glorious pink. Dark clouds thread this nebula. In the lower right corner, an immense cloud of dust is a haven for molecules. In 1978 an entirely new star began to switch on in Orion, thus proving that it is a cosmic nursery.

The Milky Way

On a clear dark night stars spangle the sky, looking near enough to touch. In fact, most of the stars visible to the naked eye are within a thousand light years. Apart from the twinkling stars, the faint band of light called the Milky Way stretches across the heavens. This silvery haze is tens of thousands of light years away. With field glasses, or a small telescope, the Milky Way is revealed as a dense crowd of countless thousands of faint stars. Within the band of the Milky Way, the combined light of millions of remote stars is just sufficient to be visible.

Since the Milky Way forms a complete circle around the heavens, parts of it are visible from everywhere on Earth, along with such important constellations through which it passes as Cassiopeia, Perseus, Auriga, Monoceros, Vela, Crux, Scorpius, Sagittarius, and Cygnus. The richest star fields lie in the southern Milky Way, forming a splendid sight in South America, Australasia and southern Africa. For northern obser-

The Milky Way is similar to a vast pinwheel. The Sun is located toward one edge, thirty thousand light years from the center. Open star clusters are found in the main disk. In a halo surrounding the galaxy are a hundred or so globular clusters, which were formed about fourteen billion years ago.

vers, the Milky Way is at its finest late on summer evenings when Cygnus, the Swan, is overhead.

But what precisely is the Milky Way? It is the view, from the inside, of the great starry Galaxy to which the Sun belongs. This Galaxy has perhaps one hundred billion stars altogether, and because it is observed from within, it is not easy to visualize the shape directly. In fact, the Milky Way Galaxy is shaped like a disk, with a bulge in the center, a gigantic whirlpool with starry arms that wind around the central part several times. The distance across the Galaxy is 100,000 light years. From Earth it takes 30,000 light years for a radio message to travel to the center of the Galaxy. And to count all the stars, one by one, would take a thousand years at the rate of three each second.

The brightest part of the Milky Way is in Sagittarius, and from this region radio and infrared telescopes detect strong signals. At the chaotic center of the Galaxy there may even be a black hole, somewhere in the direction of Sagittarius, that is gobbling up gas and possibly stars and planets that drift in too close.

The Galaxy is rotating around its central regions, but it does not turn like a solid wheel. Stars near the center make one orbit in only a few million years, but out near the Sun a single circuit takes 250 million years. Since its formation, the Sun has only managed twenty round trips, and since people have appeared on Earth the Sun has made less than one-hundredth of a galactic orbit.

The slow wheeling of the Galaxy, with the inner sections outracing the outskirts, means that the stars themselves are steadily drifting about the sky. In thousands of years the constellations will look quite different.

In this wide-angle photograph of the northern sky, the Milky Way spans the heavens. The hazy light of the Milky Way is due to the combined effect of a host of remote stars. The Galaxy – named the Milky Way because of its whitish appearance from Earth – is a vast family of one hundred billion stars held together by the force of gravity.

The Galaxies

The Milky Way is huge, but beyond it stretches a whole Universe that is unimaginably vast. Throughout this space there are scattered countless millions of other galaxies, some like the Milky Way system, and others completely different. Just as the Sun is only one among the myriads of ordinary stars, so the Milky Way is just one ordinary member of the universal family of millions of galaxies.

Astronomers did not even realize that there were galaxies outside the Milky Way until two centuries ago, when William Herschel made his famous survey of the sky. His telescopes kept bringing misty nebulas into view. Some of these, Herschel reasoned, must be Island Universes, far outside our own system of stars. In the early twentieth century careful studies of variable stars confirmed Herschel's ideas. The nearest large galaxies are millions of light years distant from us.

Galaxies come in a number of shapes and sizes. Mostly they are either spiral, like the Milky Way Galaxy, or elliptical, something like a lemon. A small proportion is irregular in shape; these misfits seem to have been disturbed in some way. The number of stars within a galaxy may vary from a few hundred million up to perhaps a trillion in a giant elliptical galaxy. The Milky Way is a large galaxy both as regards size and number of stars.

At one time it was thought that elliptical galaxies gradually became flatter and then sprouted spiral arms. This is completely wrong: there are two main types of galaxy that are made in different ways and neither can turn into something else.

Apart from visible stars, the spiral galaxies also contain gas and dust. When spirals are turned edgewise toward the Earth, as some of them are, dense clouds of dust can be seen in the spiral arms. Elliptical galaxies do not have much gas: it either turned into stars or was blown away when the galaxies formed.

The galaxy M83 in the constellation of Hydra is a spiral that faces Earth wide open. Brilliant young stars and glowing gas clouds stand out like gleaming jewels in the spiral arms of this galaxy, eight million light years away.

The two nearest galaxies are the Large and Small Magellanic Clouds, named in honor of the great sixteenth-century Portuguese navigator Ferdinand Magellan, who noticed them during his round-the-world voyage. They can be seen from the southern hemisphere as large hazes of light. Both of these irregular clouds of stars are dwarf galaxies, respectively 160,000 and 190,000 light years away.

The great galaxy beyond the constellation of Andromeda is similar to the Milky Way. It is about two million light years out in space and is the most remote object visible with the naked eye. Light now visible from Andromeda started its journey to the Earth before human beings existed. Yet telescopes can now photograph galaxies that are three billion light years distant. A radio message sent to any creatures who may exist in these far-off galaxies would take six billion years to arrive and for the reply to be received. Clearly humans will never be able to converse with other galaxies. What astronomers can do is to study them with telescopes of all types in order to accumulate as much information as possible. Theorists can then attempt to understand the physical processes which originally formed the galaxies.

Andromeda (above), a magnificent spiral galaxy, is two million light years from the Milky Way. Andromeda has two smaller elliptical galaxies in orbit round it.

The Large Magellanic Cloud is a dwarf galaxy about 160,000 light years distant from the Earth. This image in red light shows many of the galaxy's gas clouds as well as its star clusters.

Exploding Galaxies

Radio astronomers made the first discoveries of very active galaxies in the 1940s. A fascinating example of a radio galaxy lies in the constellation Centaurus. Detective work done on Centaurus A showed that the radio waves were coming from a huge galaxy twelve million light years away. On each side of this galaxy are two clouds of electrically charged particles. These particles crash through a magnetic field, emitting strong radio signals. Each radio cloud is far bigger than the Milky Way. The highly energetic clouds were probably thrown out of the Centaurus galaxy in an explosion.

Radio-emitting galaxies shine out like beacons in the remote parts of the Universe. In fact, radio telescopes detect the strongest radio galaxies at distances that are well beyond the reach of present optical telescopes. Studies of the visible light sent out by the nearer radio galaxies confirm the idea that they have turbulent central regions, rocked from time to time by explosions.

In some of the active galaxies the radio-emitting clouds have been thrown hundreds of thousands of light years into the voids between galaxies. In others, they are still confined to the central part of the optically visible galaxy. Several of the most powerful active galaxies are, additionally, strong X-ray sources.

What kind of energy is powering these astonishing radio galaxies? After thirty years of thought scientists have an outline of what is happening. Somehow an immense black hole, millions of times bigger than the Sun, is formed in the center of a galaxy. Matter – stars, planets, and gas – gets sucked into the hole and is partly changed into energy. The energy released is great enough to eject much of the central region of the galaxy, sending a flood of electric particles far beyond the galaxy.

Apart from the radio galaxies, astronomers have observed other varieties of turmoil between the galaxies. Just as the Earth and Moon are linked by the force of gravity, so the galaxies sometimes get locked together in pairs. Then tidal forces rip their spiral arms to shreds. Long streamers of stars and gas may get stranded in empty space.

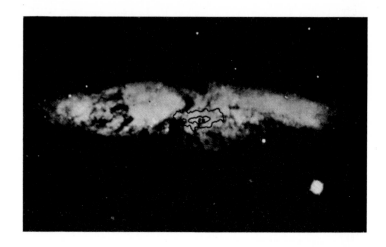

Galaxy M82, shown above, was noted in the eighteenth century by the French comet hunter Charles Messier. This object is considered to be a peculiar galaxy. It is thought to be unusually rich in dust and gas, which form billowy clouds, and to be an active site of star formation.

The photograph below shows a small cluster of four galaxies locked together by gravity. Gravitational tidal forces, similar to those that cause the tides on Earth, have distorted the shape of the lower pair of galaxies.

The galaxy Perseus A, or NGC 1275, was one of the first sources of radio waves from beyond the Galaxy discovered by astronomers. The tangled filaments may be threads of gas in intergalactic space that are being drawn into Perseus A.

Centaurus A is a beautiful radio galaxy in southern skies. It is crossed by a band of thick dust. Outside this galaxy are two enormous invisible clouds that act as strong emitters of radio waves. Perhaps Centaurus A is powered by a black hole millions of times more massive than the Sun.

The Centers of Galaxies

When observers followed up the discovery of powerful radio galaxies, they stumbled across a mystery. Some sources of radio waves could not be matched to visible galaxies. Instead, the optical telescopes detected tiny points of light in the locations of some radio sources. Although, at first glance, the objects looked like stars, spectra showed that these "radio stars" were far stranger than any star within our Galaxy.

These intriguing and distant objects became known as quasars, an abbreviation of their original name, "quasistellar radio sources." The main properties of a powerful quasar are these: it sends out a great flood of light, as much as a hundred or even a thousand giant galaxies. This energy flows from a region of the quasar that is much smaller than a galaxy, which is why a quasar looks, on a photograph, as tiny as a star; and the spectrum shows that the quasar is billions of light years away. Many quasars emit radio waves; a few send infrared, ultraviolet and X-rays as well.

Quasars may be the most distant objects ever seen. Although on ordinary photographs they look like stars, the spectrum of light from a quasar tells a different story. Lines in their spectra are shifted toward the red end of the spectrum. Most astronomers think this happens because the quasars are traveling at high speeds, somewhere from one-tenth to nine-tenths the speed of light. These high velocities show that quasars are very remote objects. Indeed, the remotest quasars are so distant that the light that astronomers observe was emitted when the Universe was less than one-third of its present age. Their light had already made half of its journey to astronomers' telescopes when the Sun and Earth came into existence. So the light from these far-off objects comes to us almost from the beginning of time.

Radio, optical and X-ray telescopes have shown that the central powerhouse within a quasar is only about as big as the solar system. Yet within this tiny volume, the quasar pours out as much light as billions of stars. How can it do so?

The answer is not clear. Disturbed and active galaxies, however, give us certain clues. At least some of these abnormal galaxies have unusually large amounts of energy gushing from their central parts, or nuclei. Astronomers believe that a giant black hole in the center of the galaxy is responsible for this strange activity. Material crashing down into this central black

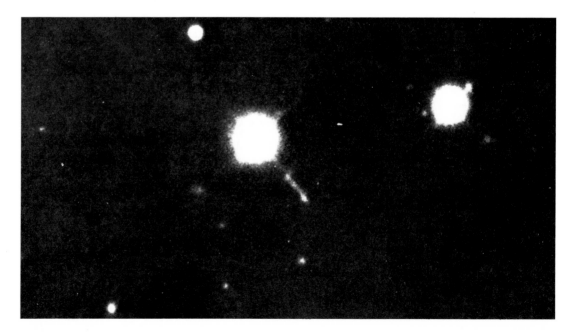

Quasar 3C273 (center) has blasted out a jet of electrons from its central regions. The electron jet emits an eerie blue light. Within other quasars, jets of matter have been expelled in a similar way. The matter speeding away from a quasar may have a velocity as high as half the speed of light relative to the quasar.

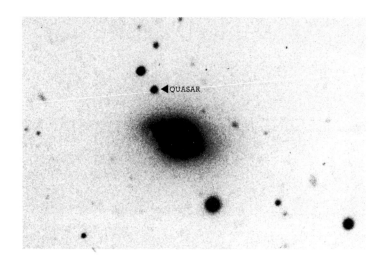

A quasar that apparently lies close to a lemon-shaped galaxy looks like a star (all the other images in this picture are stars) rather than a galaxy. Measurements show, however, that the quasar is very much farther away than the galaxy. This quasar sends out as much light as dozens of galaxies.

The sequence of pictures below summarizes possible stages in the life of a quasar: at left, the black hole at its center releases energy to furnish the quasar's intense light; at center, the black hole has consumed all the stars within reach; at right, the quasar has evolved into an ordinary spiral galaxy.

hole releases the energy detected by telescopes as a quasar.

The suggestion is that quasars may be the intensely active centers of very distant galaxies. An ordinary galaxy at the same distance as a quasar is too faint to be glimpsed through our telescopes. But the disturbed nucleus of an abnormal galaxy is like a beacon, much brighter than the underlying galaxy itself, and only this central part is visible.

If the energy source of a quasar is thought to be an extremely massive black hole it must contain the remains of billions of stars in a collapsed region of space merely a few miles in diameter. Only a black hole so massive can make enough energy to power a quasar. Exploding stars and nuclear catastrophes on a cosmic scale cannot liberate energy fast enough to keep a quasar burning. Only gravity, the force that ultimately controls the Universe, can unlock the vast energy supply that a quasar needs, as the black hole voraciously "swallows" stars.

Many of the quasars were formed very early in the history of the Universe. They are visible today not as they are now, but as they appeared long, long ago. By now they may have burned out and possibly transformed themselves into normal galaxies. If this idea is indeed correct, perhaps some of the galaxies near us, even the Milky Way, are dead quasars. Even the Milky Way may have a small black hole, of less than one million solar masses, at its center.

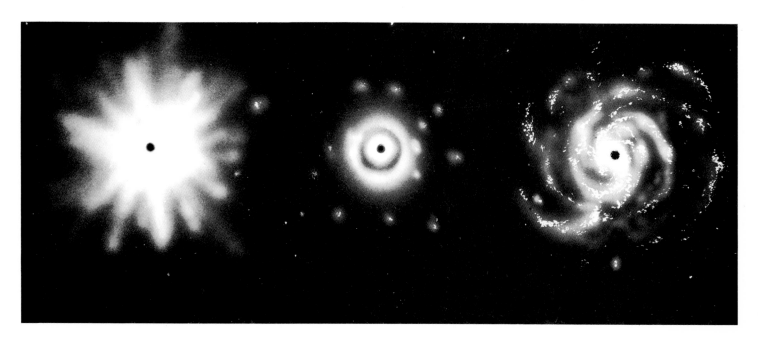

The Origin of the Universe

Humans live in the midst of incomprehensible immensity. In every direction teeming galaxies crowd the telescopes, and radio signals flood in from the remotest quasars. How and when and where did this all begin?

Cosmology is the branch of astronomy that deals with the properties of the Universe as a whole. Until a few years ago this was a task that only brilliant mathematicians could work on, using formulas and equations to deduce the nature of the Universe. Now the situation has changed. Many observations tell about the Universe and its history.

An important discovery relates the distance of a galaxy from the Earth to its observed speed. All the faraway objects are apparently racing away from the Earth. The farthest ones are moving fastest. At a distance of one billion light years the galaxies are rushing out at one-tenth the speed of light. Some of the quasars are speeding off at three-quarters the velocity of light. The curious fact that objects a long way off are traveling out at high speeds, shows that the Universe is expanding. Retracing the paths of the oldest speeding galaxies, it appears that they were all in one place between 10 and 20 billion years ago.

The Earth is not at the center of the expansion, however. No matter where an observer was in the Universe, he would see everything else rushing away from his galaxy as fast as possible.

As astronomers now believe, the early phase of the Universe, sometimes called the Big Bang, that flung out the galaxies, was unimaginably hot, so hot that even today the entire Universe is glowing with the after-effects. Radio astronomers have found that space is filled with feeble radio waves. These are literally the dying radio echoes of the Big Bang. The radio waves show us that space is not totally cold. Rather, the heat from the Big Bang has given the whole Universe a temperature that is 5 Fahrenheit degrees above absolute zero ($-459°F$).

Careful measurements of the speeds of the galaxies show that the Universe as we know it has existed for 15 billion years. Astronomers cannot say what happened before then. Perhaps the Universe emerged from the collapse of an earlier Universe, but no one can ever know for certain if this is the case. The Earth, Moon and Sun have existed for little more than one-third of the time since the origin of the Universe. The birth of civilization, the foundation of towns and the organization of human society have taken up less than one-millionth of the age of the Universe. What will mankind accomplish in the next millionth? Can the nearest stars be colonized?

In ten thousand years' time there might be human colonies around local stars. But it is hard to imagine how the Universe itself can be colonized. Telescopes have already detected quasars that are ten billion light years away. Billions of galaxies sprawl through the Universe, most of them billions of light years away. The sheer size of the cosmos is a great barrier to communication and travel. Science can only learn about the Universe at large distances and early times by studying the wheeling galaxies and flaring quasars. That is part of the ageless fascination of astronomy, the oldest of the sciences, now passing through one of the most exciting eras in its long history.

As reconstructed in four phases by an artist, no galaxies existed in the initial burst of the Big Bang (top left) – only fragments of atoms and immense heat and light. After a million years the first threads of hydrogen gas formed (top right) from the atomic particles. The gas then clumped into cloudy masses greater than entire clusters of galaxies (bottom left). Finally, the galaxies themselves condensed into clumps about fifteen billion years ago (bottom right). Today the galaxies are far apart and the Universe is thinly populated with stars and galaxies.

Glossary

Asteroid: one of the numerous small rocky objects (also known as minor planets) that orbit the Sun in the solar system

Atmosphere: the layers of gas around a planet, a natural satellite or star

Aurora: luminous curtains or streamers of light sometimes seen in the night sky at high northerly or southerly latitudes, also known as the "northern lights" or "southern lights"

Big Bang: a theory for the origin of the Universe which suggests that all matter and energy in the Universe was once concentrated into an unimaginably dense state, from which it has been expanding since an event between 10 billion and 20 billion years ago

Binary star: a pair of stars in orbit around a balancing point between them, held together by the gravitational pull that each has on the other

Black hole: a region of space where the strength of the gravitational pull exerted by the matter concentrated there is so powerful that even light cannot escape

Chromosphere: part of the outer layers of the Sun, visible as a thin crescent of pink light in the few seconds immediately before and after a total solar eclipse

Cluster: general term to indicate a group of stars or of galaxies sufficiently close to one another to be physically associated

Comet: a diffuse object, made of gas and dust, orbiting the Sun in the solar system, and noted for the spectacular tail it develops on approaching the Sun

Constellation: one of the eighty-eight named areas of the sky, which together cover the whole of the sky's sphere. The term is also used to describe the pattern formed by the brightest stars within the bounds of this area

Corona: the outermost layers of the Sun, visible as a faint halo around it during a total eclipse

Double star: a pair of stars that look close together in the sky. They may either be a true binary pair, near each other in space, or they may just happen to lie along the same line of sight.

Eclipse: the total or partial disappearance from view of an astronomical body when it passes directly behind another object; or the passage of a moon or planet through the shadow cast by another body so that it cannot shine, as it usually does, by reflected light

Fraunhofer lines: narrow, dark lines cutting across the continuous, rainbow-like spectrum of the Sun, and of other stars

Galaxy: a giant family of stars, held together by the gravitational pull exerted on one another

Gravitation: one of the four natural forces, always attractive, and acting between all objects with any mass, but decreasing rapidly with the distance between the objects

Infrared radiation: a radiation similar in nature to light and radio waves, but with a wavelength intermediate between the two

Interstellar matter: gas and dust between the stars

Light year: a unit of distance used in astronomy, defined as the distance traveled in one year by light. It is equivalent to almost six trillion miles.

Magnetic field: region of influence around a magnetized object, and the pattern and strength of the magnetism within that area

Magnetosphere: region surrounding a planet in which the planetary magnetic field rather than the solar magnetic field controls the motion of electrically charged particles

Magnitude: a measure of the brightness of a star or other astronomical object

Mare: Latin word meaning "sea," which is the name used for the dark plains on the Moon, believed to be solidified seas of lava

Meteor: the momentary streak of light seen in the sky as a fragment of interplanetary material burns up while speeding through the Earth's upper atmosphere

Meteorite: the remains of a larger piece of interplanetary matter that survives passage through the atmosphere and falls to the Earth's surface

Meteoroid: a fragment of rock in interplanetary space that could become either a meteor or a meteorite

Moon: a popular term for the natural satellites of any planet. With a capital letter, it refers to the Earth's own moon

Nebula: Latin word for "cloud," applied to any objects in the sky that appear hazy and do not have a sharp edge. It is used more strictly to mean true clouds of gas, and clouds between the stars

Neutron star: a tiny star, a trillion times more dense than the Sun, primarily made of the atomic particles called neutrons

Nova: a star that brightens suddenly by as much as eighteen magnitudes, then gradually fades over a period of months

Orbit: path followed by an object moving under the action of a gravitational force

Parallax: the way that objects at different distances from an observer appear to change their relative positions as the observer moves. Parallax provides an important method of measuring distances.

Phase: the portion of the disk of the Moon or of a planet that can be seen by reflected sunlight

Planet: one of the larger, non-luminous bodies orbiting the Sun in the solar system, or a similar object that might orbit another star

Planetary nebula: a shell of glowing gas, surrounding an old star from which the gas has been ejected in the past

Prominence: a streamer of glowing gas visible in the outer layers of the Sun

Pulsar: a radio source that sends out regular bursts of radio waves at intervals of a few seconds or less. Pulsars have been identified with neutron stars.

Quasar: an astronomical object of starlike appearance whose spectrum shows peculiarities suggesting that it is moving away from the Earth at a considerable fraction of the speed of light

Radiant: the point of the sky from which meteor tracks belonging to a family of meteoroids diverge

Rille: elongated depression in the lunar surface

Satellite: a small body, either natural or man-made, in orbit around a larger body

Solar flare: a sudden and short-lived brightening in the Sun's atmosphere, caused by a discharge of energetic particles

Solar system: the Sun and its associated family of planets, satellites, comets and other interplanetary matter

Solar wind: a moving stream of electrically-charged particles from the Sun

Star: a ball of material held together by its own gravity, and generating radiant energy that makes it shine

Sunspot: an area on the Sun's surface that appears slightly darker than surrounding regions because it is at a slightly lower temperature

Supernova: a rare and spectacular stellar explosion that results in the destruction of a massive star, leaving behind a neutron star

Ultraviolet radiation: a radiation similar to light, but of a shorter wavelength, often emitted by very hot objects

Van Allen belts: two zones encircling the Earth in which electrically charged particles are trapped

Variable star: a star whose light output is not constant

White dwarf: a small, dim star that is close to the end of its life

X-rays: a radiation, similar in nature to light, but of a very much shorter wavelength and great penetrating power

Index

Credits

The publishers gratefully acknowledge
permission to reproduce the following
illustrations:

Anglo Australian Observatory 7, 27t, 70, 81, 87b;
© Association of Universities for Research in
Astronomy Inc. The Cerro-Tololo Inter-
American Observatory, The Kitt Peak National
Observatory 62, 64, 65, 76r, 83, 86, 89, 90;
J. Allan Cash Ltd. 9; Dennis di Cicco 34, 35;
Georg Gerster/John Hillelson Agency 13; Hale
Observatories 76l, 80, 87t, 88, 91; Gary Ladd
27b; Lick Observatory 32/3, 39; Lowell
Observatory 46; Courtesy of Martin Marietta
Aerospace 47t; Denis Milon 50b; Jacqueline
and Simon Mitton 31, 51b, 66; Mount Wilson and
Las Campanas Observatories, Carnegie
Institution of Washington 79t; Mount Wilson and
Palomar Observatories 63; NASA 28b;
29 (David Baker); 43l, 49b (Space Frontiers);
30, 43r, 45t, 49t, 68b (Science Photo Library);
NASA-Ames/David Baker 45b; NASA-Jet
Propulsion Laboratory 28t, 47b, 48, 50t, 61l;
3, 52, 53, 55, 56, 57 (David Baker); NASA-World
Data Center 36, 37, 40, 41; Ondrejov
Observatory 85; Royal Astronomical Society 60,
61r, 71, 79b; Royal Observatory, Edinburgh 82;
Sacramento Peak Observatory 67; Science
Photo Library 26 (D. Parker), 68t (Dr. L. Golub),
69 (J. Finch); John S. Shelton 51t; TASS 44;
University of Michigan 73; University of Toronto
75.

Cover photograph: NASA-Jet Propulsion
Laboratory/David Baker

Artwork by: Fiona Almeleh 10, 14; Gerard
Browne 8, 25, 31, 42, 79; Aubrey Dewar 6;
Carol Kane 15, 18/19b, 22/3b; Carol McCleeve
23, 32, 34, 67, 74t, 75, 78; Michael Robinson 7, 13,
72/3; Rob Shone 11; Mike Strickland 9, 12, 16–17,
18t, 19t, 20/1, 22t, 33, 54, 63, 71, 74b; Ian West
77, 91; Tim White 93.

Bibliography

Cambridge Encyclopaedia of Astronomy,
    Simon Mitton (ed), Crown Publishers Inc.,
    1977
Catalog of the Universe, P. Murdin and D. Allen,
    Crown Publishers Inc. 1980
Galaxies and Quasars, W. J. Kaufmann, W. H.
    Freeman, 1979
The New Solar System, J. K. Beatty and
    B. O'Leary, A. Chaikin (eds), Cambridge
    University Press, 1981
The Search for Life in the Universe, D. Goldsmith
    and T. Owen, Benjamin/Cummings, 1980